UID Journal Workbook

Volume 1

Jim Bensman

Copyright © 2008 by Jim Bensman.

Library of Congress Control Number:	2008901432
ISBN:	Hardcover	978-1-4363-2313-0
	Softcover	978-1-4363-2312-3

All rights reserved. No part of this book may be reproduced or transmitted in any form or by any means, electronic or mechanical, including photocopying, recording, or by any information storage and retrieval system, without permission in writing from the copyright owner.

This book was printed in the United States of America.

To order additional copies of this book, contact:
Xlibris Corporation
1-888-795-4274
www.Xlibris.com
Orders@Xlibris.com

45350

Contents

Preface 9
Foreword 11
Acknowledgements 15
Introduction 17

Chapter One: Implementation Strategy 25
Chapter Two: Scanners 101 27
Chapter Three: Enter the Matrix 31
Chapter Four: The Matrix Speaks 36
Chapter Five: A Horse Of A Different Color 41
Chapter Six: Reporting The Data 44
Chapter Seven: Reporting The Data, Part Two 49
Chapter Eight: UID and RFID 55
Chapter Nine: UID + RFID + AIT= Win + Win 61
Chapter Ten: Time Marches On: A Million Parts Later............. 65
Chapter Eleven: UID PMO Policy Issues/Information 70
Chapter Twelve: UID Implementation and Contract Retention 74
Chapter Thirteen: Effects of UID Implementation
    On Manufacturing WIP... 76
Chapter Fourteen: Validation, Verification and Grading............ 79
Chapter Fifteen: Direct Part Marking 101 83
Chapter Sixteen: From Offense to Defense: RFID At The DoD 89
Chapter Seventeen: Keeping Track of Other People's Property......... 92
Chapter Eighteen: Flawless Medical Asset Tracking................... 97
Chapter Nineteen: UID For 'Newbies' 100
Chapter Twenty: UID For Newbies—Advanced 106
Chapter Twenty-One: Selecting a Systems Integration Firm ... 112
Chapter Twenty-Two: Ask Dr. UID 114

The AIDC field that we find ourselves in today is one of the fastest paced development areas of modern business. UIDUK have spent years refining their systems and expertise to ensure when advice is given it is cost effective, tried and tested. I have seen IUID and RFID solutions deployed all over the world helping enrich every level of our technological existence. In a world that is becoming smaller every day and information that is hungrily sort out by every age group; the AIDC technologies deliver every time, on time and with accuracy. There will be no area of our daily lives that will not be touched by or supported by AIDC technology. As a board member of AIMUK I am privileged to see these new technologies first hand, and watch them evolve into the market place changing both business practices and working life within the defence and public sectors. In this book we will talk about many aspects of UID deployment and the processes that will ensure you will have a structured approach to this fascinating field of AIDC. The book has been written by James (Jim) Bensman who is one of the foremost experts in the field of UID. I have worked with Jim both in the UK and in the States and he has become a good friend fuelling my desire to push the boundaries of all these technologies and ultimately forming UIDUK.

Peter Young
Director of Operations—UIDUK.
Board member—AIMUK
Committee Member—UID Supplier Alliance

WHAT THE EXPERTS ARE SAYING ABOUT UID JOURNAL WORKBOOK, Vol. 1

". . . If you are already in a project or just starting on one, the UID Journal Workbook is one of those books that once read will sit by your right hand. Its straight forward delivery of information and advice makes a must for anyone working in this field . . . "

Peter Young
Director of Operations
UIDUK

"The UID Journal Workbook is an invaluable guide to walk suppliers to the DoD and military depots through the intricacies of UID. The Step-by-Step process aids both neophytes and experienced users in understanding what needd to be done to insure full UID compliance."

Peter Ginkel
VP Business Development/UID
ID Integration, Inc.

". . . The UID Journal Workbook provides the fundamentals and detailed lessons to assist organizations achieve UID compliance. By utilizing MIL-STD-130 requirements . . . you are able to realize the full benefit of UID to achieve internal profitability and efficiency."

Freedom Technologies Corporation

"I have been working in the UID world for several years developing both UID CD training courses and university programs relating to UID. I have reviewed the UID Journal Workbook and found it very comprehensive. I strongly recommend this book to anyone getting started in a UID project or someone who needs additional insight into UID."

Robert W. Rylander and Associates

Preface

By Rudy Pearce,
Managing Director UID UK

UID UK who are we?

UID UK is a marketing brand of GSM Graphic Arts Ltd, who for the last 30 years have been the leading manufacturer of industrial labels and nameplates in the UK. As part of the GSM Group we provide labels and marking solutions in to all market sectors.

The **UID UK** brand covers all manner of asset marking and tracking but we specialise in Item Unique Identification (IUID). We have worked very closely with the North American Supplier Alliance (NASA) in the USA to fully understand the requirements and implication of the US Government Mandate to code all the assets of the Department of Defence (DoD). **UID UK** have a member on the board of AIM UK who are responsible for the development of new coding technology through out the UK, we are also member of NASA which enables us to keep right up to date on what is happening State side.

We know you will find this 'Workbook', which was penned by Jim Bensman, one of the foremost authorities on UID in the States extremely useful in understanding both the principles of UID and the practical application of it.

Foreword

By Richard A. Erickson
Program Manager
Lockheed—Martin I-GUIDES

As taxpayers, we all want to see the Department of Defense, (DoD) use our money to provide the right assets at the right time. This is especially true in wartime when a battle can be determined by the logistics support or more importantly a life put at risk. Since 2003 when the DoD mandated certain products be delivered with a two dimensional ECC200 compliant Data Matrix, industry has been scrambling to become compliant and employ this technology. It will allow the DoD to drastically improve asset visibility for the entire service life of a product. When contractors employ this technology as part of their internal operations, it improves internal operating profits.

Employing this technology is no longer an option. This Workbook will be indispensable as we adopt the use of this technology. It captures the basics of the technology from leading industry experts as well as outlines opportunities for integrated applications. The Workbook brings a full spectrum of opportunities for you to consider within your business environment.

As one would expect with a new mandate and new technologies being released across the DoD community, changes to policies and governing specifications will continue to occur. Changes in the DoD input schema are also occurring. Without this Workbook to outline the DoD policies and reference materials, it would be difficult to keep current.

The section on Scanners will illustrate some of the challenges you will face in the selection of hardware to read the 2D matrixes as well as a lively discussion on just how to optimize the potential inherent in the system. The Workbook gives you pointers on how to achieve maximum results. The section on RFID alone is worth the investment in both time and money not only in understanding RFID and it's potential but it provides an invaluable primer on *why* the DoD is so interested in implementing it world wide.

Our experience in providing this technology and working with the DoD supply base is that each customer has their own set of unique requirements. The good news is that industry is responsive and very capable of providing solutions that meet or exceed individual requirements.

At the early IUID Forums sponsored by the Office of the Under Secretary of Defense—Acquisition Transformation and Logistics OUSD/ATL, we had conversations with suppliers who needed to permanently mark helicopter blades, parachutes, and well drilling pipes. These are tough technology challenges and the industry groups that addressed these challenges have contributed to the knowledge base collected in this Workbook. It will be important for you to reap the benefit of their experiences as you begin your journey. It is also important to review the wide range of software and hardware components that are available from the supplier community.

Exploring this Workbook prior to beginning an implementation of the IUID and or RFID technologies will provide you and your organization with a variety of approaches that others have used to deploy these technologies. The Workbook also describes the efficiency expected once an integrated solution has been employed.

Since incremental technologies like IUID and Passive RFID have become mandated, DoD providers have sought to integrate these disparate technologies to address the larger challenge of what the DoD calls "Utilization". Utilization attempts to respond to the statement—"it's all about the data".

Now that you have developed or captured the correct unique identifier data and assigned it a unique item identification (UII)—what do you do with that "birth record?" Since the mark is intended to be permanent and unchanging throughout the life of the product, data associations from the birth record to vital supply and support data elements can be established. Organizations can now establish maintenance records and harvest them for vital life cycle operational information to make decisions for any part of the theater of operations. The example of using the data to extract "bad actors" from the supply support system is seen as a huge cost savings. Discussions by industry innovators and the pioneers of problem-solving that address this large role one the IUID is established are contained in this Workbook.

Other technologies covered in the Workbook include Auto Identification Technologies (AIT). AITs are touched on in several of the chapters in this Workbook. When your organization employs these technologies to capture data it reduces or eliminates human intervention and manual data collection. Significant returns on your investments can be achieved. In studies and pilot programs completed over the past years this has been substantiated repeatedly.

One such pilot program reduced the time to develop and use data by a factor of 7.5! When this factor is translated into costs savings, error reduction and error correction, the investment dollars spent to install the IUID and RFID infrastructure is soon recovered.

Regardless of your role in making IUID a vital part of your business, this Workbook will serve as an invaluable reference document and stay in demand for the foreseeable future. I encourage you to embrace the use of this technology and enjoy the many resources this Workbook brings to you.

January 2008

Richard A. Erickson
Program Manager
Lockheed—Martin I-GUIDES

UID Journal Workbook, Volume 1 Acknowledgements

The author is grateful to many individuals in industry and business, a few of which are mentioned here.

A special expression of gratitude to Ms. LeAntha Sumpter UID Program Manager as well as Mr. Robert Leibrandt, the UIDPMO Help Desk, and Mr. James Clark. Special thanks to the UID Journal staff, especially Mark W. Lee, Sr. Technical Editor for UID Journal.

Special thanks to Richard A. Erickson Program Manager at I-Guides, Lockheed Martin Maritime Systems and Sensors. Also, to Andrew Cary and Jane Yallum Clarion Communications Group for their ongoing support and encouragement. A debt of gratitude to Dr. Tom Edison of DAU for insightful support and encouragement. Retired Lt. Col. USAF Greg Redick, for his invaluable ability to understand the nature of this complex issue and for his kind encouragement along the way.

The author wishes to thank Gary Moe, and the entire staff of ID Integration, as well as Rick Scorey and the team at Freedom Technologies for their help and encouragement. The author also wishes to express gratitude to Matt Van Bogart, Product Manager at Microscan Technologies, as well as Ron Caines of Psion Teklogix.

The author is responsible for content and is grateful for assistance with technical accuracy as provided by Mr. Robert Rylander of Rylander and Associates. And to Mike Faber of The Idea Works for his tireless dedication to seeing this project through to completion.

Introduction

By LeAntha Sumpter
Deputy Director for Program Development and Implementation
Defense Procurement & Acquisition Policy
Office of the Under Secretary for Defense for AT&L
UID & eBusiness Forum Sept 12th 2007 Atlanta, Welcome Address & Status Update on IUID & eBusiness Policy & Strategic Plans

What is so unique about identification that we need Unique ID? How have we been managing parts and have we been managing the information about parts forever? The answer is, of course we have. What is so unique about identifiers that we need something called UID?

Let's take a look at the problem first. The challenge that we have, in an organization as a large and diverse as Department of Defense (DoD), is that we have a lot of brilliant people who have developed business processes that are unique across the entire corporation, we have so much diversity that we don't have is enough commonality around data management so that we can make sense out of our data.

A number of years ago we took a long look at this problem and two people—Mike Wynne at that point Deputy Under Secretary for AT & L, and General Cartwright, put their heads together. There actually was an SROC (Senior Readiness Oversight Council) guidance that was issued in 2002. It said that we need a set of data keys for DoD. We cannot make the mass of data that the department has, relevant in the operational space, unless, we can use a set of data keys that will bridge our functional community worldwide and supercede individual native systems. So that was the foundational basis—the guidance that came out of SROC . . . and one of their goals was the Unique ID of Items.

What that was projected to do is it sets up the foundation for a common AIT (Automated Information Technology) infrastructure. It also allows us to integrate communities (departments and services) that right now are different. As an example we were able have a look at the finance community and the logistics community—both of which have different processes, different structures, different forms, and different data elements. We order by DoDACC's and we manage property by CAGE (Commercial and Government Entity #). Those two distinct universes really do not intersect too frequently.

When we looked at the Maintenance community, we were surprised. Actually the Maintenance community is, at least within ATA (Air Transport Association) a very high performing community that has always had a burning need to have integrated processes for purposes of aid certification, safety items and all of the processes that really are world class, in which DoD really is a world class leader.

Could they be improved? Yes.

Essentially, what we are doing is working with the communities, figuring out exactly what we need to do to improve. Where we are today? Where do we want to be tomorrow? We don't need marks today, at the end of the day, what we need is a set of data keys and permanent, unique identification is that set of data keys. Its all driven by and going back to the SROC guidance. Why? Data keys are manageable on a DoD scale.

We also have the take it to the next level, so if you look at the functional list of processes is one component of this; the part-marking piece is another component. Then you have to look at how we manage the data about parts.

What we do is we have a lot of vocabularies—we have to be able to talk to everyone and make sense of it all in the end. We have parts, and items that are uniquely tracked. We serialize them today. We mark them and track them. We mark them, and we **don't** track them. We track them, and we don't mark them. And, my favorite-that's the one where you don't have a unique serial mark or ID on that part, but we manage it that way. I don't know how we do, do that today in a very, very labor-intensive environment. That is not efficient and effective, in terms of the corporation.

Then, there are those items where we don't mark them, and we don't track them, but we should be. There are those items where we don't mark them and we don't track them, nor should we-at the item level. We manage them and track them at the class level. And that is absolutely a program, and the only thing that makes business sense and we should continue to do it that way.

What we are trying to do is to look at the end of the process. What is the data we want to manage and be able to see it for logistics, for accountability, for accounting, for procurement, for acquisitions, for program managers, for life cycle—I need one key. That's what UID is about. You can see how large the problem is and how simple the concept of the solution is. The devil is in the execution.

So why UID?

What is so great and special about UID?
About 2002, we locked, what I still consider to be—25 to 50—of the smartest thinkers and doers in the world, we had people from the U.K., from industry, whiz kids from the automotive industry—we literally locked them in a room along with some of the brightest minds from the standards committee at DoD.

I believe God intervened to help us because it also snowed. We couldn't get out and no one could get into this hotel we were meeting in. What they did is come up with the UID Structure. That group of 50.

Frankly, what I discovered is, I learned there is a small group of motivated individuals in the world who are passionate about this stuff. Most have been in the business for the better part of 30+ years. They look at how you manage parts, how you manage data, about how a system is structured—there is a lot of depth, thought and a lot of history to this and at how to manage data in a maintenance environment, a logistics environment and an acquisitions environment. They came up with the UID Structure. It was truly a collaborative solution.

What it did was bridge multiple standards organizations, but it was built on the premise that what we needed was a common infrastructure for a Machine Readable Language about parts. But where we were at that point is we had 300 standards communities that frankly, had had a divorce. We had ATA—standards environments used by the airlines, which also includes most of the large DoD contractors, if they are involved in the aviation business; we had ISO (International Standards Organization); and then we had the UCC community-which is consumable goods, and frankly, those three Machine Readable Languages did not interoperate.

The backbone to the UID Structure is a couple of things. One is: We de-centralized parts management. We said; if you put your enterprise ID in front of that part mark, on that part, and we need a serial number and a part number-you control your serialization. We are not going to control it. You are controlling it. Completely de-centralized.

The other thing we did is broker an agreement between those standards bodies so that they would recognize each other's Machine Readable Language. Which then created the environment so that the DoD *internationally* could move to a standard environment for management of parts. This spans everything we do.

Why 2-D?

2-D (2-D data matrix) was invented by NASA. It was created so we could mark small stuff. Linear bar code doesn't work because it took up a lot of space on small parts. We needed a mark that could work, that was reliable and frankly 2-D was the next generation part mark. DoD actually had proposed the standard in the early 90's but had no traction until 2-D came along.

Another part of why this had been a break within the DoD standards community was that this was where they had wanted to go 10 years earlier.

Is it a commercial standard? Yes. In fact, most the world-class manufacturers we talked to were already using this construct—Japan and other countries were way, way out there with it. Many European countries also were using this. It's fascinating that the further you get into this, how many folks from outside the US are ahead of us in terms of how they are going to manage data holistically system wide using a common key, using a 2-D data matrix instead of a linear bar code.

The only thing a part mark does, is it keys up the opportunity to manage the data with less space, with less chaos, in terms of personnel. One of the things the Navy looked at was how to use UID to manage Surface to Surface missiles which are installed to protect the country worldwide, that have a 100% inspection criteria every year. What they discovered was they save 96% of the personnel cost of managing that process. The payback was in less than six months. It's "All About the Data".

We need a much more efficient process so that what we do is put our money into supporting our troops, instead of managing our data-space. We think we have at least 97 million items within the UID Space. What we need to have is a much more efficient and effective process for managing uniquely serialized items.

What is the policy framework for this? That is another part of the complexity. This is not just about procurement, or acquisitions, or finance, or logistics—it's about all of the above.

In March of 2007 there was a DoD directive that we have set keys of which IUID is one of them. Why did we do that? Because that is what our guidance was when we started in 2002.

The second key that just came out was a memo by Dr. Fenway, directing that we have to be able to integrate government serialization and UID together in order to get what we need, to go for maximum capitalization of assets, as well as managing maintenance. What does that mean? In addition to UID, we need the UNO #, the tail #, the bumper #-you know the numbers that we control, whether on the hull of a ship or the tail of a plane? Those also have to integrate with UID. So the (UID) Registry has been modified to allow us to integrate those two data elements together, making them relevant as aliases, so that we get them to work in multiple disciplines at the same time.

The next piece is an instruction in DoD that assigns responsibility for managing policy and also establishes what the authoritative responsibilities are for the IUID Registry-and that is that it is the authoritative source for acquisition value of items acquired after Jan.1st 2004 and the pedigree data of delivery.

New rule Published Sept. 13 2007.

It also establishes it as the master source for property. On September 13, 2007 an interim rule was published on GFP, which has a brand new contract provision mandatory for all contracts recorded after September 13, 2007 that makes the UID Registry the single point where all GFP (government furnished property) information is managed. The other thing is we are working on is a national-level, data-strategy for traceability of information. We are going to be able to look across the enterprise at the functional level and manage it holistically, which is really difficult to get your head around.

The other thing we are still working in the standards committee, to get a formal blessing from ISO allowing that common construction to co-exist, so that you can use the machine readable *language* and *syntax* whether you are operating in the aviation environment, the GS-1 environment or you are operating in ISO. One month after we declared our adoption (several years ago) there was a meeting of 21 engine manufacturers community worldwide—why have multiple standards? So, to meet the engine manufacturing community, worldwide, clearly this is a huge detail.

There are a lot of standards pieces, but we still need the part mark MIL-STD for UID, bringing in the part-mark and RFID data (on the packaging), bringing in all this data synchronously. We are also still working on another iteration of MIL-STD-130 that is the part-mark MIL-STD for DoD.

We are also working on a NATO STANAG as well as a NATO implementation policy. There are an awful lot of companies/countries in Europe that are looking at us, waiting to see when DoD makes the nickel drop. The power for this in the coalition space is HUGE, if we all migrate to a single data standard.

We are working to design a warranty track. DFARs needs to track warranty across the line, across the lifecycle. The interesting thing is we just reached out to services and to give us examples of the variety of clauses that they use in the warranty space. We received a stack of pages three inches thick on the warranty clauses from the Air force alone. This is the diversity in the warranty space. The goal is to create data for the operational forces, for our logistics community and for procurement community because we cannot put a contract out, until we develop the warranty issues.

There is also a DFAR's case out on the Wide Area Work Flow (WAWF). The WAWF is the ONLY system in DoD in which we are going to do receipts against an invoice. We are hoping to use the business rules we have already built with DLA as they began to roll out in defense that establishes the collaborative relationship between what the ERP does, and what the WAWF does. The good news is that the Navy has finally decided to deploy big time

Its not the big guys—the big companies, big guys don't have as much trouble getting their act together with UID. Lockheed Martin is the composite of 75 legacy companies-each of those companies had a parts management culture that they brought with it, as well as a parts management structure within an engineering environment and a logistics environment-do you think this is easy for them? But when we approached them four or five years ago, do you know whom we talked to? The first entrée we made was with their systems engineering side. Their advice to me was that this requirement really has to be a finance initiative, not just an engineering initiative, even though engineering might be a big fit. So the CFO of Lockheed Martin is leading the transformation as a partner with the logistics side of Lockheed Martin. To become truly compliant takes a multi-disciplinary arrangement; this is not anything that can be led from any individual discipline area.

This is the grand capstone of the directive I mentioned earlier, at the end of the day the DoD sends people and stuff to do a mission. Very Logistically. What we need to do is to operate in a coalition state and operate where what we do is we budget and we manage in what I call . . . organization but we deploy this way—we take people from all different force within the government. It could be Task force Charlie, and guys from the CIA, some people from SAIC, and an army/navy mix, and throw a couple of men from SOCOM in there and all have to do a mission with parts from God knows where. How do you track that, how do you manage that? Well, you've got to have a couple of keys which is the genesis of that directive I mentioned back in 2002. So what leadership decided was we needed to have Unique Identification for Real Property, Personal property. That's your CAC card (Common Access Card), Organizations, and Location. Every single one of those now are going to fit in a deployment, so that what has taken us from 2002 until now to actually put a definition around each of those. And we can now integrate that data in a meaningful fashion.

So what we done is come up with a set of phases:

1) Real Property (GFP) 07
2) Military Equipment Valuation slated completion next year (08); integration the GFP piece and the Real Property piece—actually a Navy program
3) Integrating Acquisition Programs and the organizational foundational piece.
4) Moving beyond that, we will determine what else can be done" with UID

LeAntha Sumpter
Deputy Director for Program Development and Implementation
Defense Procurement & Acquisition Policy
Office of the Under Secretary for Defense for AT&L
UID & eBusiness Forum Sept 12th 2007 Atlanta, Welcome Address & Status Update on IUID & eBusiness Policy & Strategic Plans

Chapter 1

Implementation Strategy

S upplier Implementation Strategies

Many successful companies continue to bid and win progressively larger projects, while meeting delivery dates and production demand.

The result: Congratulations! Your company has received a contract award from the Dept. of Defense. However, many companies are noticing contractual language with which they may not have previously been familiar: WAWF, Mil STD 129 and UID.

Companies may undergo stress when they have to suddenly plan for an implementation that may not have been fully anticipated. Perhaps management had put the brakes on or simply planned improperly—something that often happens with newbies. How do Suppliers and contractors address issues of compliance with DoD mandates?

In many companies, an implementation committee is formed consisting of a team of representatives from vital departments, to address the UID implementation project and resolve crucial details: Contract, Warehouse, Shipping, Manufacturing, IT, Accounting, Engineering, and other affected departments.

The *UID Journal Workbook, Volume 1* is designed to assist companies with identification of relevant issues and anticipate the repercussions of the solution implementation upon company processes. Most companies with high volume production and automated processes can simply not afford to interfere with that schedule.

UID Journal adds insight to:

- How to automate the process, meet the requirements and maintain our production schedule?

- What information is needed and how is it associated with the contract number and line items (CLINs) then sent along with ASN (Advance Shipping Notice) to the DoD using WAWF (Wide area workflow)?

  - How is the UID 2-D data matrix applied to the product?

  - How many UID products per carton; cartons per pallet?

  - How do we associate UID components within the system UID?

  - Pallets and cartons may require RFID labels (radio frequency identification tags). How do we associate UID data to the RFID label?

  - UID Journal adds insight to How UID data is associated with the RFID case and pallet labels?

  - What about mixed pallets, and random box shipments?

  - Does Data from ERP (Enterprise Resource Planning) drive the ASN?

  - What if a contracting officer has indicated that we may have to prove the validity of the UID Marking process by quality reports grading each UID 2-D data matrix, how do we accomplish this?

  - What is the cost of implementation/training of our UID Solution?

Chapter 2

Scanners 101

Part One:
Of Systems and Sensors

Resisting the natural urge to delve right into the fun stuff—those nifty, state-of-art devices we get to hold in our hands, wave around and acquire information as if by magic—let's linger for a moment at the systems analysis level. Boring perhaps but ultimately necessary to keep things in perspective.

Every system comes into existence for a reason and UID systems are no different. Though you may suspect that the unstated goal is to torture suppliers, there is an official reason—Compliance.

Of course, external agents—in this case, DFARS and DoD, impose compliance from the outside. So right from the start we are all fighting an up-hill battle, encountering organizational resistance, swimming against the current—however you want to put it. Countering this inertia is some motivation, expressed here in the most primal terms by the draconian phrase "contract cancellation" and its evil sidekick "loss (or delay) of revenue." Take heart, though, because this effort can also establish a solid foundation for improved manufacturing processes inside your company.

From an operational viewpoint, the functions that have to be implemented to meet government compliance can be clearly stated:

Marking
Verification
Validation
Registration

On Your Mark

Though included as one part of the overall system, Marking is a science unto itself and important enough to be considered as a separate realm. After all, without a good mark, there's nothing for a sensor to sense and no data for the system to handle. Also, these guys have to go first and cannot escape dealing with the practicalities of the physical world. According, there is a wealth of solutions offered for marking which can be broken down as follows:

Traditional
- Pin
- Ink Jet
- Laser
- Data Plates
- Alternative
- Electric Arc Spray
- Epoxy
- Spray and Fuse
- Emerging
- RFID

The material being marked and the requirements often dictate choice of the appropriate solution for permanency. Data plates offer versatility because they have the luxury of controlling the substrate material and are prevalent enough to warrant their own break down:

Substrates: Paper, Polyester, Polyimide, Aluminum, and Stainless Steel
Printing Methods

Surface: Laser toner, Ink jet, Thermal transfer

Altered Surface: Chemical or Laser etching, Impact, Laser bonding or marking

Integrated: Photographic

Each of these solutions claims advantages in specific applications and we refer you to the literature provided by the manufacturers of marking systems for a lively discussion concerning these tradeoffs.

Get set

So now we have the means for applying the mark, but what of the mark itself? What does it contain? Information in the form of numbers and codes is the answer, which is only of benefit when understood throughout the system. The encoding of this information is standardized by the DOD and given the slightly redundant name of Item Unique Identification (IUID). In simplest form this consists of an Enterprise Identifier code (assigned by ISO) combined with a serial number. If the serial number is not unique within the enterprise, then the original part number is also included.

Scan

Now that we have something of import to scan we can get down to business. Getting back to primary system functions, the first use of your scanner is to verify that the mark has been applied correctly. Verification also creates the initial data record in the system and includes preliminary information such as manufacturer, serial number and an indicator of how well the mark can be read.

Many things can go wrong when creating or scanning a mark, so much so that standards have been established to set the parameters for this evaluation. For 2D bar codes the standard is ISO/IEC 15415, which deals with concerns such as distortion, non-uniformity, pattern damage, low contrast and resolution.

Of primary interest is the Unused Error Correction Capacity indicator. Because software algorithms can correct for damaged or poorly modulated marks some indication is needed as to how much of this correction is being applied. Otherwise, problems with the original marking system can be masked and you would lose the opportunity to correct the marking system. 100% unused error correction is ideal and indicates no damage to the mark.

Validation can be viewed as a form of re-verification. Typically this is a second scan done at another station, at which time additional information is appended to make the data record complete. Then the data record is uploaded to the central UID Registry for subsequent use through out the supported infrastructure.

Thanks for your patience, but I couldn't bear going into the particulars of scanners without setting the context of their use in the overall system. Anyway, we'll address the intricacies of sensing technologies next chapter when we ask:

'Hey Joe, where you goin' with that scanner in your hand?'

Chapter 3

Enter the Matrix . . .

PART TWO
Enter the Matrix . . .
Scanners 101, Part 2

OK, you patiently endured the last chapter's basics explaining why you've been given a nifty hand-held scanner and how its use is part of a serious, important system aimed at collecting revenues, etcetera, etcetera. So, go head, squeeze the trigger—scan your mark.

Done already? My, photons are phast. But what just happened, right before your very eyes? Certainly a number was produced—in fact, a highly unique number, pertaining only to that item and all its clones. This probably explains that certain psychological boost you feel while scanning an object. You know, that sense of preciseness, which can be expressed through a slight snapping of the wrist. It's sort of like stamping something, only in reverse. As if to say, "there, nailed it."

But what did the scanner actually "see". Well, certainly not the same thing our eyes see—which is the source of both great benefit and even greater hassles.

Benefits: not clear to the eye

Years ago the demand to cram more information into smaller spaces forced everyone beyond human-readable marks. Think here of a small bolt and the requirement to have its unique identifying number appear on its head.

Unlike angels dancing on the head of pin, representing characters on the bolt can't help but take up *some* physical space. You can go with smaller lettering, but you're quickly into the realm of microscopy—which may retain the benefits of immediate human interpretation but it also introduces an external device. So, we might as well use optical technology then to encode characters into some scheme that maximizes density. Now you've crossed over into the realm of machine-readable identifiers (MRI).

Barcodes are certainly the most prevalent forms of MRI marks and successfully meet the needs for a host of applications, but the simplest forms are one-dimensional (1D) which establishes the limit as to how much information (i.e. characters) can be represented in a given space.

Adoption of a two-dimensional (2D) approach increases this upper limit exponentially. These 2D marks consist of black and white cells arranged in either a square (typical) or rectangular (for oddly-shaped items) pattern called a Data Matrix. Favored by the Electronics Industries Alliance (EIA), the data matrix now in the public domain and has been codified into an ISO/IEC standard (16022) thusly:

Using this approach (and current technology) as many as 50 characters can be represented in a 2 mm square.

Of course, this is an ideal image and seldom attainable—which is not necessarily an issue because, software algorithms can be applied to the original image to extract this ideal image. So this graphic also illustrates what the scanner *wants* to see—in the same way (I suppose) that The Ring *wanted* to be found in the classic Tolkien trilogy: *The Lord of the Rings*.

Anyway, assuming all went well, what the scanner actually 'saw' before processing was something similar to this image from a dot peened mark:

A bit daunting isn't it? Some wit stated it was like going from "Map" view to "Satellite" view—isn't it? Still, this can be made to work. But what if all *did not* go well? What if you can't even capture an image this good? Then, welcome to the real world.

Hassles: *as though one is looking through a glass darkly* . . .

Let's get back to photons, because that's ultimately what we are dealing with here. That is, whether the sensor inside the scanner you're holding is laser-based or a CCD, it's picking up reflected photons. (In a future chapter we'll deal with RFID.)

So, first you have to optimally illuminate the target area in order to even locate the mark. This is where the applications engineers get involved because you can't treat a 2D mark the same as a 1D mark—and hope to succeed. Uniformity across the mark is what you're after, with the end goal being the creation of high contrast image. To attain this, much work is done with how light is applied to the mark. Important parameters include the angle of the illumination (in relation to an axis running from the object to the sensor) and how diffuse the lighting is made. The material and its coatings determine which combination works best.

Ultimately, illumination is driven towards either a 'bright field' (dark dots on a light background) or, inversely, a 'dark field' (light dots on a dark background) solution. To the software, these images are equivalent because it is looking for transitions either way. What can't be tolerated is when the same image from a given mark contains both bright field and dark field areas—which can easily happen, for instance, using direct lighting for a dot-peened mark on a polished metal rod. So, variations in diffuse and off-axis lighting are used to eliminate contrast reversals within an image.

OK, you've done all you can do to apply a good mark and illuminate it, now it's time to turn the captured image over to the software—good luck. Ultimately, we know the software is going to produce a number code, but first it must locate this code.

To facilitate this initial—and therefore difficult task—the specification calls out a 'quiet zone' surrounding the code, which is wider than a cell and devoid of any patterns that might be interpreted as bit transitions.

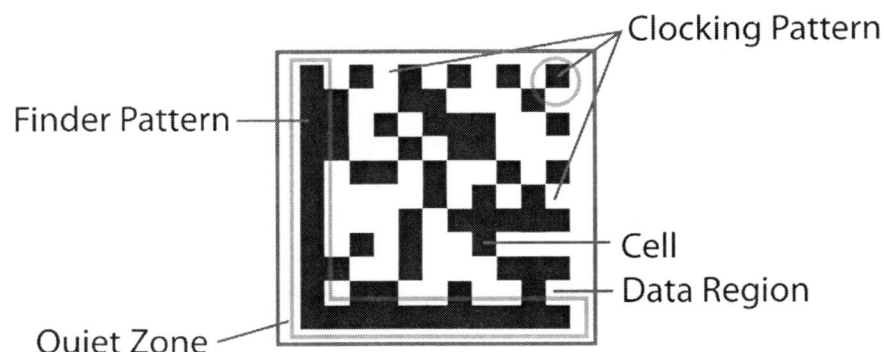

Once this zone is identified, then the software can start looking for the clocking pattern and the finder pattern in the matrix (see below). This is similar to the silent 'gaps' used in serialized data transmissions which allow the reader algorithms to 'sync up' and begin looking for the actual data.

Similarly, knowing the cell size and locating the finder pattern, enables the software to begin extracting the data encrypted into the matrix—which now that all the preliminaries are out of the way was our original goal. Besides numerical part information, the encoded data matrix also includes error-correcting codes (ECCs) that allow algorithms to correct—within limits—for damaged or poorly imaged marks.

The importance of ECC cannot be overstated. It can recover matrix marks that are more than 50% damaged, but more importantly it relaxes practical system tolerances such that less-costly technologies can be employed. However, it's important to keep in mind that error correction cannot be applied until the matrix is located and decrypted. Which means ECCs cannot be used to recover a damaged finder pattern, so they are of no benefit during initial locating of the mark. According, in practical application, the emphasis continues on the quality of the initial mark and the illumination during subsequent readings.

Next Chapter: ***The Matrix Speaks or Secrets of the Code Revealed*** . . .

Chapter 4
The Matrix Speaks

The Basics
The Matrix Speaks

Prepare now to leave the mundane world of physical concerns such as parts marking and image scanning, for we now enter the ephemeral realm of data communications. Alas, the matrix has been made to babble and we must now interpret what comes forth.

To begin you must come to see the symbol for what it really is. Just like an artist's palette and canvas, it is—in the end—merely the *medium* for the all-important *message*.

So let's reveal the matrix symbol for what it truly is—a data carrier. That is, the medium selected to record, transport or communicates data (American Heritage Dictionary). The data that it's transporting is information about a particular item that's already available inside the organization that's producing it—part number, lot number, and serial number. Only now it's all funneled down and encoded into a symbol according to an agreed-upon set of rules, so anyone with the right equipment can subsequently decode this information.

What's the point of all this, anyway?

The functional goal here is Item Unique Identification—IUID when reduced to its DOD acronym. That's the goal; the way to achieve this goal is to create a "set of data" for each "tangible asset" that is "globally unique and unambiguous". (I just hope I'm not being ambiguous here, because that's exactly the sort of thing we're supposed to be stomping out.) At the very core of this whole scheme is this "set of data" and it's given the moniker Unique Item Identifier (UII) or more generally, the Unique Identifier (UID).

To create this unique identifier, all available and necessary information about the part and the organization producing it is linked together—concatenated—into a string of text characters whose ultimate destination is a 78-character data field awaiting it inside the UID Registry that "ensures uniqueness" of each item listed in the registry. Of course, along the way this unambiguous identifier also sees prime use within the manufacturing organization for its own tracking and control purposes

That's the goal all neatly stated, but how is this achieved in practical application? As usual, the devil is in the details—and for those we return to the physical world and your scanner. Once the matrix symbol is imaged and decoded it offers up a string of characters.

Inside the String - courtesy Mark Reboulet, AF AIT PM

Record Separator · DUNS Data · Part Number DI · Serial Number DI · End Of Transmission

[)> R_S 06 G_S 12V 194532636 G_S 1PA234 G_S S786B50 R_S $E_O{}_T$

Format Code (DIs) · Group Separator · DUNS DI · Group Separator · Part Number Data · Serial Number Data

Notice that this string contains both 'printable' and 'non-printable' characters. The inclusion of non-printable characters enables the software to first isolate and then interpret the data as it goes by in a stream.

Algorithmically and paradoxically, you start by looking for the "End of Transmission" character. This is one of many non-printable standardized (ASCII) 'control' characters—which date back to prehistoric times, when teletypes ruled the world. Next a pre-ordained (ISO 15434) start string is searched for—and once identified the software is said to be "synced up" and you can begin detailed interpretation of the string's content.

You may find it useful to think of these preliminaries as analogous to the UPS person delivering a package—they assures its safe transport, but are not directly involved with its contents. Extending the analogy, the remaining control characters can be viewed as defining the package, itself. In our case, we catch a small break because each package contains only one record. However, each record (package) is subdivided into groups.

At last we can open our package and view its contents. First, notice that the start of the record contains a format code—which means different *types* of records, can be supported. The format code indicates the structure of the groups within in the record that follows. Three formats are acceptable:

DI: containing Data Identifiers
AI: containing Application Identifiers
TEI: containing Text Element Identifiers

Why the three types? And which one should you choose?

We went to the source: James Clark from the UID Help Desk explains: "Over time different industries have developed different semantic formats for encoding data into barcodes. For IUID purposes, there are three formats that are acceptable: AIs, DIs, and TEIs. These were selected because of their common use between defense and aerospace industries, and have been developed and managed through international standards bodies.

"The first question to ask is whether or not your company currently uses one of the data qualifiers formats—if so, use that one.

"For example, if you are in the aerospace industry, you probably use TEIs, which are the accepted data qualifiers within the aerospace industry, are governed by the Air Transport Association standards and are listed in their Common Support Data Dictionary.

"DIs are maintained and governed by American National Standard (ANS) MH 10.8.2, while AIs are developed and governed by GS1 (formerly EAN-UCC). So if you are a member of GS1, more than likely AIs are the way to go.

"If you do not currently use any of the formats, then you can choose any of the three. You may want to review the different standards and standards organizations that govern these data qualifiers to determine if any of them make more sense than the others, otherwise, you are free to choose any of them."

Whichever scheme is chosen, each group in the format begins with a Data Qualifier code that indicates specifically the information contained in that group. Consulting the following table that shows what the three formats look like and how they relate to each other, you'll find familiar items like serial number, part number and, most important to this discussion, the Unique Identifier (UID).

One bit of finesse appearing in each format is the last entry for *Current* part number—as opposed to the *Original* part number. This supports the reality of part numbers rolling as time goes by.

Though you may be suspicious, the complexity reflected in this table really is necessary because different organizations need to convey different types of information. Even if they share the same type of information, it is often available in different forms. An important example should suffice:

Sometimes a serial number can be kept unique across the organization (think title plates); other times the serial number is only unique for that part. Consulting the table below, you'll find that either situation can be supported, regardless of the format used.

When an item's serial number is unique across the company (Enterprise), then the UID is termed Construct 1; otherwise the serial number is considered to be unique only within the part number and is termed Construct 2. Accordingly, the part number is additionally included in the unique identifier field for Construct 2.

The information about which construct is being used to create the UID is so important in practical application that a separate field is reserved for it in the Registry.

	UII Construct #1	UII Construct #2	
Based on current enterprise configurations	If items are serialized within the Enterprise	If items are serialized within Part, Lot or Batch Number	
UII is derived by concatenating the data elements IN ORDER:	Issuing Agency Code Enterprise ID Serial Number	Issuing Agency Code* Enterprise ID	
		Original Part# Serial Number	Lot or Batch # Serial Number
Data Identified on Assets Not Part of the UII (Separate Identifier)	Current Part Number**	Current Part Number**	

* The Issuing Agency Code (IAC) represents the registration authority that issued the enterprise identifier (e.g., Dun and Bradstreet, EAN.UCC). The IAC can be derived from the data qualifier for the enterprise identifier and does not need to be marked on the item.

** In instances where the original part number changes with new configurations (also known as part number roll), the current part number may be included on the item as a separate data element for traceability purposes.

Once again, what's the point?

That reminds me, it's time to return our attention to that scanner still in your hand. After slogging through all this detail, you have the luxury of not having to keep it all in your head. That's the job of your scanner and its associated software—to acquire the image, parse the string, interpret the data and—voila—produce the Unique Identifier. So scan away!

Chapter 5

A Horse of a Different Colour

UID Meets RFID

Although we know that the part is marked with a UID, we learn that the carton and pallet that the parts are shipped in receive RFID Labels that electronically associate the UID contained therein.

In the last issue we generalized matrix symbols as a type of a data carrier—pedantically defined as "a medium selected to record, transport or communicate data." Then we proceeded to outline how this data carrier functioned within the material supply chain to provide the fundamental "globally unique and unambiguous" identifier as required by the DoD.

This issue our focus shifts—moves down the electromagnetic spectrum, if you will—to another form of data carrier. This is one that functions as a complementary workhorse within the same system—only it does so at different frequency (or color if we only possessed the appropriate sensors).

Instead of using photons, this carrier traffics in electrons. Instead of capturing an image with light, information is collected through radio waves. It is precisely this shifting in the electromagnetic spectrum that makes Radio Frequency Identification (RFID) useful as a supplemental technology.

In situations where optical scanning is impossible or disadvantageous, RFID can be put to use instead. One of these areas—and one that is prime interest to the DoD—is packaging. In the eyes of the DoD the use of RFID technology to tag an item's packaging has distinct advantages mainly because it is not a 'line-of-sight' technology.

Bar codes and matrix symbols (actually two-dimensional bar codes) must be 'seen' by a scanner in order to be read. The person who functions to first locate the symbol so it can then be scanned, of course, operates the scanner. Which is not to say that optical scanning is never automated, especially when it comes to part marking; but the DoD cites RFID as having inherent, multiple advantages for the scanning of packages—mainly because "human intervention" is not required to capture an RFID tag.

Moreover, optical tags must be scanned individually and care must be taken to only scan each optical tag once. RFID tags, on the other hand, include a unique digital identifier, so RFID readers so *multiple tags* can be scanned at the same time *and* tags can be scanned several times but only record the item once. This means that an entire shipment of goods labeled with RFID tags can be moved through an RFID reader quicker and that all items in the shipment are identified only once—eliminating multiple item capture errors.

In general, interest of DoD in RFID for packaging runs deep because it views RFID to be well suited to its overall goal of attaining Automatic Identification Technology (AIT) throughout the material supply chain. Much of this is because an RFID tag can also be written to. That is matrix symbols can be generalized as "write once, read many"; whereas RFID tags are "'write many, read many." Storage capacity of RFID varies from miniscule to volumes of data, and cost of the RFID tag scales accordingly, especially when compared to matrix symbols.

Once cost justified, data within a tag can provide increased levels of information for an item during manufacture, transit, storage, use, or maintenance. With this additional data, the tag can support applications that need item-specific information. For example, shipment consignee or destination ports can be readily accessed upon reading the tag. The DoD lists other such applications.

Getting back to our tortured analogy, saddlebags have now been added to our horse (data carrier) allowing people to include notes of interest to certain other people along the way. In DoD speak this provides *enhanced item level visibility*—which addresses a key short-coming that the DOD has noted as occurring at every node along its supply chain.

So, by way of review, when it comes to packaging the main attractions of RFID and their ancillary benefits can be summarized are as follows:

- No human intervention
- Less error prone
- Able to read tags faster
- Better suited to automation
- Ability to read multiple tags
- Better suited for packaged items
- Eliminates duplicate scan errors
- Ability to write to tags
- Allows addition information
- Enhanced item level visibility

Once again, we must stress that all this is not a case of either/or—not optical scanners verses RFID readers—but one in which *both* technologies are used to assure successful identification and tracking of material throughout the material supply chain.

Cleary, the long-term view of the DoD is that the use of RFID, just like part marking, not only solves problems of its own concern throughout the material supply chain, but also enables a supplier to improve its own internal processes; so the DoD's requirements for immediate implementation and future expansion are appropriate. That's the proverbial stick; the carrot is that suppliers will get paid quicker.

So, with that in mind, let's turn out attention to the practical application of RFID (in the next chapter).

Chapter 6
Reporting the Data

Give Data, Get Paid.

In this chapter, we're going to tell you the secret of how to turn information already in your possession *into real money!*

In truth, it's not really a secret—at least it's not intended to be—but to the uninitiated the whole process can be intimidating.

- To get a running start, let's summarize the discussions from previous issues into the following actions:
- Mark the part and its packaging
- Verify that the part and packaging can be correctly read
- Validate that the mark creates an globally Unique Identifier (UID)
- Register the UID and its associated data with the DoD

So, the ultimate destination for all your work is the UID Registry maintained by the Defense Logistics Information Service (DLIS). Into this registry you will be placing not only the (UID) but also a whole raft of data associated with it—and the part it represents. When completed this Item Master list contains more than 30 items organized into the following categories of information:

In subsequent usage, the descriptive and contract categories are often grouped together and called the Pedigree information. This becomes important when you want to get paid.

UID Journal Workbook

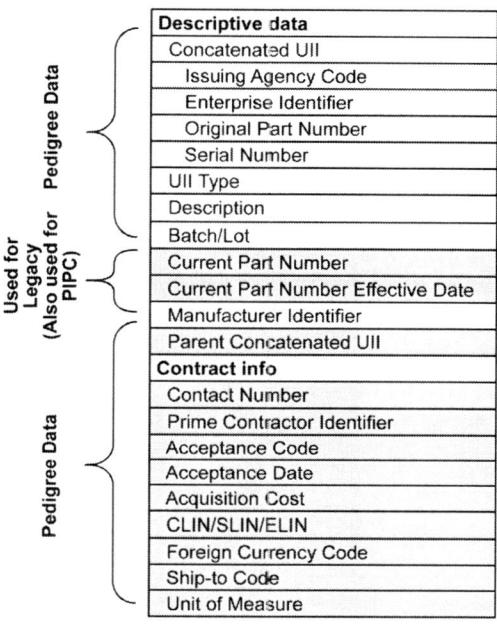

Now we're ready to discuss the exact mechanism for getting your precious information into the UID Registry. To do this you'll be using the Wide Area Work Flow (WAWF)—which means we can longer avoid The Big Picture:

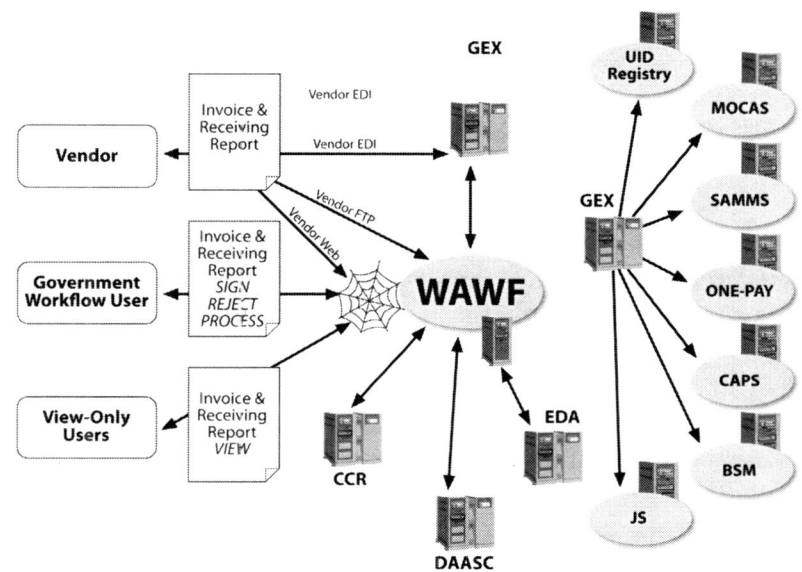

At the functional level, the WAWF is a transaction-based system created by the DoD to support the procurement lifecycle from submission of electronic invoices through inspection, acceptance and, ultimately, electronic payment receipt. For the DoD, the WAWF is pivotal in realizing its stated vision regarding Government Furnished Property (GFP):

> *"DoD, its coalition partners and industry will efficiently and effectively manage individual tangible items of Government property with near real-time situational awareness using globally unique item identification, enabled by speedy and accurate automatic data capture".*

This vision is then translated into the following key principles:

Create data once, use often

An UID Registry will:

- Maintain Master UID data
- Be updated with key transaction events
- Maintain transaction history
- The UID in the Registry will be common throughout the WAWF
- Acquisition value will only be recorded and updated in the UID Registry
- Accounting will be separate from accountability meaning that the UID registry will never become the accountability system, but will be the audit trail of current and previous accountability systems.

Accordingly, the UID registry is shown as a separate entity in the diagram above. More important to our discussion are the functions that the WAWF provides for the Registry. That is, the WAWF documents the transfer, shipment, receipt or acceptance of new acquisition items or existing GFP by the DoD and contractors. It does this by capturing transaction data and submitting the appropriate information to the UID Registry.

In implementation, WAWF is a secure web-based system for electronic invoicing, receipt and acceptance. It creates a virtual folder to combine the three documents required to pay a vendor—the Contract, the Invoice, and the Receiving Report. Interactive web-based applications within the WAWF enable the electronic submission of invoices, government inspection forms and acceptance documents to support DoD's goal of moving to a paperless acquisition process.

In implementation, WAWF is a secure web-based system for electronic invoicing, receipt and acceptance. It creates a virtual folder to combine the three documents required to pay a vendor—the Contract, the Invoice, and the Receiving Report. Interactive web-based applications within the WAWF enable the electronic submission of invoices, government inspection forms and acceptance documents to support DoD's goal of moving to a paperless acquisition process.

At the practical level, the WAWF helps the DoD to mitigate interest penalty payments due to lost or misplaced documents and highlights vendor offered discounts. Benefits to vendor include the improved accuracy of submitted information by minimizing re-keying of data and the online accessibility of documents, especially those required for payment, potentially streamlining the payment process from weeks to days, or even hours.

Getting started with the WAWF really isn't all that imposing; here's a break down of the steps:

- Register on the WAWF webpage: *https://wawf.eb.mil/*

- Contact Defense Information Systems Agency (DISA) to request FTP access: 1-866-618-5988

- DISA will submit a ticket to the JITC (Joint Interoperability Test Command)

- JITC will reply with a welcome email and contact information at JITC

- Proceed to contact JITC with info provided

- JITC will assist in completing initial setup

- Complete and fax DD Form 2875.

- Fax: 1-801-605-7453

- Submit test e-document to your JITC contact's email

- JITC will verify test documents are in correct format

- Receive secure FTP contact information

- Submit Invoices, when ready . . .

Using secure FTP (File Transfer Protocol) is just one of three ways that invoices can be submitted into WAWF. We mention it first because it is the easiest to implement and the DOD supports it, because it does not wish to force vendors to make significant changes to their existing processes. However, vendors that already use Electronic Data Interchange (EDI) are certainly encouraged to use their EDI system to submit into the WAWF.

Either one these two methods is recommended for vendors who submit numerous invoices or have involved invoices containing many line items. For vendors that submit only a few invoices or ones that can be completed quickly, an Interactive Web Application is available. Be aware that for security reasons this web application contains a session timer; so it use may become cumbersome to use in the long run.

Whichever method is chosen, the important point is that the vendor receives timely feedback on his submitted invoice. Here is the first tangible benefit for the vendor submitting into the WAWF. If a received invoice is rejected, the vendor will have the information to quickly correct and resubmit the invoice—and only the individual incorrect data items need be addressed as opposed to re-typing the whole document. Also, the vendor can view previously submitted documents and actions taken by a government official (including name and contact information) so any follow-up action required from the vendor can be initiated immediately. All aimed at getting an EFT into the Bank.

To wrap things up, we'll end with a friendly reminder: **Once Submitted—now Committed!**

Meaning that it is the responsibility of the vendor to understand the "trigger events" that require an updating of UID Registry information. The WAWF is there to help, but the vendor must implement and continue to use data capture mechanisms and associated processes to initiate and assure that these crucial updates happen.

Chapter 7

Reporting the Data

Part 2

In the most general sense, the UID Registry is the DoD's central repository for the following crucial information:

- *What is the item's description?
    - How and when it was acquired?
- *What was the initial value of the item?
    - Who has Current custody of the item (government or contractor)?
- *How is the item marked?

Of course, leading up to this, the whole effort to mark the part according to established rules and standards was to ensure that the identifier used in the Registry will remain globally unique and permanent throughout the item's life.

Once this hook is in place, then all sorts of useful information can be hung off it.

In words of the DoD's mission statement:

The Registry captures, retains and provides current and historical data regarding uniquely identified tangible items enabling net-centric data discovery, correlation and collaboration in order to facilitate effective and efficient accountability and control of Department of Defense assets and resources in support of Department of Defense business transformation and Warfighter mission fulfillment.

Given the Registry's primacy in the DoD's asset management system, it's not surprising that a glut of important item-specific information has found a home there.

This included information can be broken down further into the following four general categories—with examples:

Pedigree
- Acquisition Contract Information
- Original Part Number
- Shipment and Delivery Information

Valuation
- Initial Acquisition Value
- Changes in Valuation

Accountability
- Contractor Custody Information
- Acceptance Data
- Custodial Contract Data

Configuration
- Embedded Items
- Item Markings
- Part Number Changes

In actual implementation the registry information for a given item becomes quite detailed and involved. The specification covering the file format runs more than 20 pages, so we'll spare you that, but for those individuals demanding more details we provide the following glimpse into the details of the parts master:

Just to let you know that you are not alone and that the Registry is a growing concern. As of summer 2006 more than 300 fellow contractors had registered more than 163,000 items. Interestingly, almost 60 percent of these contractors are small businesses. September 2007 was the deadline for all DoD *SERIALLY MANAGED* Assets registered in the Item Unique Identification (IUID) Registry!

IUID

UII	D90536RRR0012
UID Type	UID2
Issuing Agency Code	D
Enterprise Identifier	90536
Original Part Number	RRR
Serial Number	0012
Batch/Lot	
Ship-to Code	
Description	Will Test Stuff

Acquisition Contract

Contract Number	A54105-5454-2005
Prime Contractor ID	187575592
CLIN/SLIN/ELIN	A515
Acceptance Location Code	FB6222
Acceptance Date	8/1/2005
Unit of Measure	EA

Marks History

Contents	Medium	Value	Effective	Removed
PART SERIAL	HUMAN READABLE	RRR0012	08-01-2005	
UID	2D COMPLIANT	D90536RRR0012	08-01-2005	
UID	DEFINED	D90536RRR0012	08-01-2005	

Should you take a notion to visit them, the registry is maintained by the Defense Logistics Information Service (DLIS) located in Battlecreek, MI. The registry itself resides on the Integrated Acquisition Environment (IAE) Business Partner Network (BPN), so it can be accessed via the Internet. The most rudimentary method for entering the registry is to make use of the web application at *https://www.bpn.gov/iuid*.

After going through the necessary registration with the Central Contractor Registration (CCR) system, you will be able to return to the BPN system and use its web application. First though, we suggest familiarizing yourself with the 50-page Software Users Manual available at *http://www.acq.osd.mil/dpap/UID*.

Using this interactive web site for entry into the Registry is perfectly acceptable to the DoD and will probably suffice for vendors who submit only a few invoices, or have ones that can be entered quickly. However, for vendors who submit numerous or involved invoices, two other methods are available which make better use of the integrated automatic system for asset management that the DoD has put in place, known as the Wide Area Work Flow (WAWF).

As described last chapter, the WAWF is an online system created by the DoD to support its goal for electronic invoicing, receipt and acceptance. Of particular interest are its Receipts and Acceptance (WAWF-RA) extensions which provide a virtual folder and the associated workflow processes that enables the vendor, receiver, and bill payer to work together to ensure prompt payment based on electronically generated documents, while also reducing interest penalties for the government.

For vendors, the benefits of using the WAWF can be listed as follows:

- Efficient electronic submittal of documents
- Access to documents and their status 24/7
- Immediate feedback from governmental agency in case of rejection
- Real-time correction and re-submittal
- Faster processing of documents by agencies
- Secure transactions with audit trail
- No transaction fees

It's important to keep in mind that both the WAWF and Registry are integral parts of a larger system maintained by the DoD to support electronic business applications and to track electronically the creation and movement of Government Furnished Property (GFP).

At the core of this overall system is the Global Exchange Service (GEX), which provides for the translation and transportation of transactions between legacy and new systems, enabling interoperability among these systems. It combines gateway and network entry point functions into a single environment and provides an enhanced audit trail of transactions to ensure end-to-end reliability. GEX is the single trading partner hub between DoD, US Bank, Citibank, commercial shipping companies and thousands of vendors.

The relationship between the Registry, WAWF and GEX can be confusing, so the following simplified drawing of the whole system is offered:

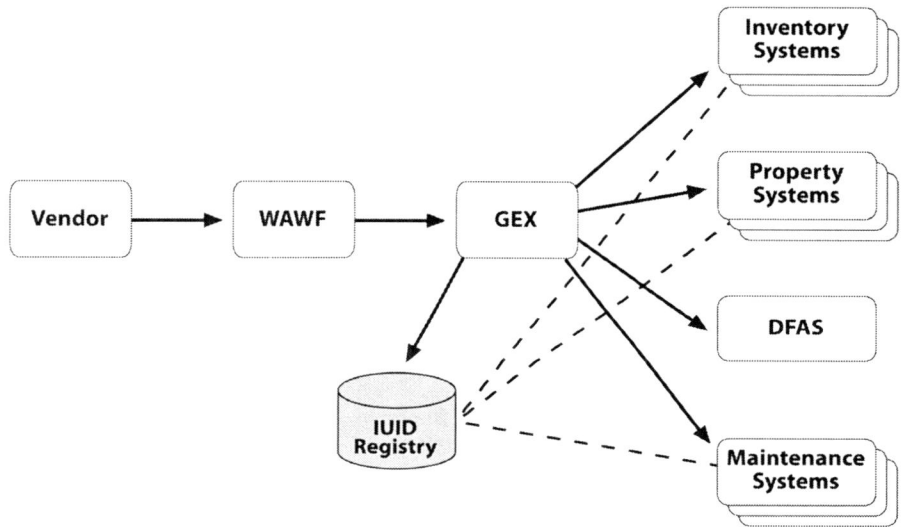

This diagram shows that the vendor makes use of WAWF, which in turn uses GEX to enter and update information for items in the Registry. More importantly, this also shows that the rest of the system for managing government property relies upon the globally unique identifier in the Registry and all the associated information for conducting its transactions and business.

Specifically, the Registry:

- Collects pedigree information on tangible items owned by the DoD.
- Provides a single reference for processing an item by assigning it's a globally unique identifier.
- Makes available the pedigree information for an item to users throughout the property management system.

While the WAWF:

- Captures transactions regarding property in the management system.
- Electronically extracts pedigree information from the vendor to the Registry.

- Processes shipment and receipt documents between contractors and DoD.

Another way to look at it is that the DoD uses the WAWF to capture transactions regarding property creation and transfers, while it uses the Registry information to coordinate all processes for a given item throughout its electronic property management system and sub-systems.

At the end of the day, the key advantage for the vendor in using the WAWF in accessing the Registry is gaining real-time and interactive access to the Defense Finance and Accounting Service (DFAS), especially its pay system and the electronic fund transfers that flow from it.

Chapter 8
UID and RFID

Interwoven Technologies, Processes, and Policies.

UID & RFID

Over the last several chapters, we followed the flow of unique identification (UID) in the DoD material supply chain from initial marking of a part through the reporting of its associated data to the UID registry ending with the final reward for the supplier of payment due. This chapter we'll revisit this same river (if you will) but from another perspective—this time concentrating on the role and use of RFID. Though the relationship between UID and RFID is in many ways complementary and their use runs in parallel throughout the DoD system, there are marked and important differences.

One way to get a handle of this relationship is keep in mind the physical object to which each of these technologies are attached: Unique identification (UID) is applied to the part—or item—itself while RFID is applied to its packaging. At first encounter this may seem like a trivial distinction, but in application and use of the information that each provides to the system this turns out to be important. One way to illustrates this is with the following diagram:

This diagram shows the multiple layers of "logistical units" in use throughout the DoD supply chain—and much can be gleaned from it.

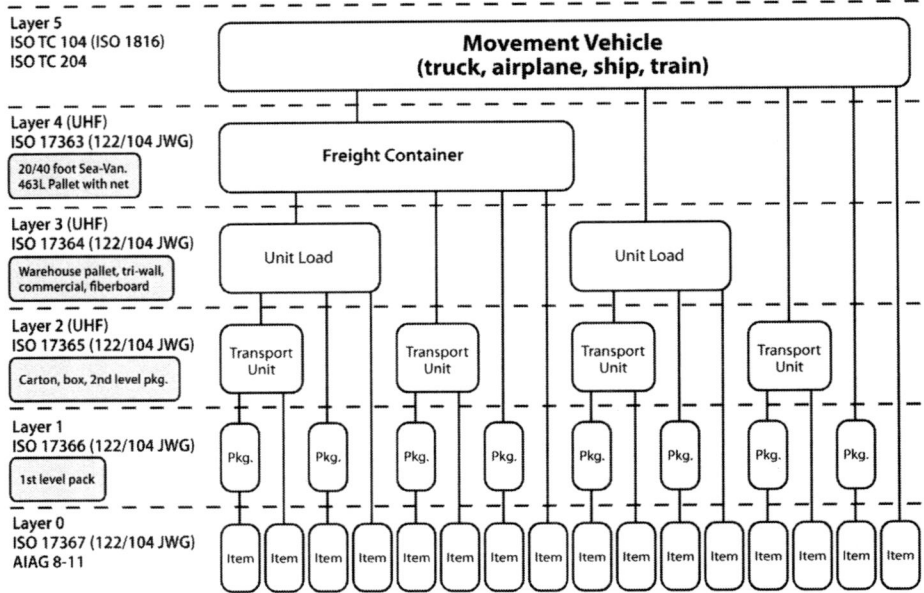

Layer 0 pertains to items and is the layer of application for UID. At this level, parts are assigned a globally unique item identifier (UII) and successfully marked with an applicable 2-dimensional matrix containing the UII. The relationship between UID and items is so strong that in discussions of the DoD supply chain you will often find the terms Item Unique Identification (IUID) and unique identification (UID) used interchangeably.

Layers 1 through 4 are the domains of application for RFID. The first important observation to be made—not surprisingly—is the transition from Layer 0 to Layer 1. Layer 0 is concerned with the physical part, while Layer 1 concentrates on its packaging.

This distinction should be kept in mind because it reflects the two policies that the DoD has put into effect for improving its supply system; the UID policy and the RFID. As discussed above (and before), the UID policy assures that individual items have unique identifiers. The RFID policy, on the other hand, relates more to the packaging with an eye towards more efficient transportation and greater in-transit visibility of the material.

Although it is only one of five levels shown, the primacy of Layer 0 should not be forgotten. Though much is done, subsequently, with associated information throughout the DoD supply system, Layer 0 is where unique identification is attached to the actual part and so, remains the foundation of the system.

However, in both of the first two levels, a form of *automatic identification* is being used. In Layer 0 it is being applied to the item itself, whereas in Layer 1 automatic identification is being applied to the item's packaging. The phrase "automatic identification" is emphasized because from the DoD's viewpoint both data matrix marks and RFID are alternate forms of Automatic Identification Technologies (**AIT**).

"RFID is a part of a larger suite of AIT applications, all of which the DoD will leverage, where appropriate, in the supply chain." states the Assistant Deputy Under Secretary of Defense Supply Chain Integration, Alan Estevez, "The data matrix is a two-dimensional barcode, an alternate form of AIT. The combination of 2D barcode and RFID technologies incorporated into AIT equipment will facilitate the UID and RFID relationship."

This relationship can get confusing, especially when use of the technologies overlap.

Notice in the diagram that the applicable ISO standards for RFID are listed for each level—including Layer 0. What's up with that? Although its UID policy requires that all parts be marked with a permanent unique identifier and its RFID policy dictates that all packaging will eventually have RFID tags, the DoD certainly does not mind if RFID tags are used on items, in addition.

Returning to the diagram, you get a good view of how material is being consolidated into larger packaging units for more efficient transportation as it moves up through the remaining layers—and that at each layer RFID is in use. However, the type of RFID being used starts to change. Passive RFID tags (un-powered) predominate for item packaging, cases and warehouse pallets, but these give way to active tags (powered) for larger pallets and SeaVan containers. Mainly this is a function of the increased read range gained for active tags. Still a connection remains with UID, with RFID technology essentially *pulling* UID information along the way.

As Estevez explains this relationship, "RFID will be used as a hands-free data collection method to identify UID items located within various levels of material packaging. In order to identify the UID item using RFID, the RFID tag data on the unit packs, shipping containers, exterior containers, and palletized unit loads must be associated to the UID information in a logistics system."

"Using RFID tags as a means of data collection and associating the tag data with UID information will help to maintain precise UID asset/in-transit visibility and to improve data quality, item management, and maintenance of UID materiel throughout the DoD supply chain. The hands-free data collection method will help extend and take advantage of the implementation of the UID policy."

One important thing to keep in mind regarding the use of RFID is that material is in reality flowing in multiple directions within the supply chain. Not only are items being deployed, but also they are being returned for maintenance and repairs or being returned to stock. All of which essentially multiplies the benefits gained from RFID tracking. The following chart gives a better overview of the improvements identified by DoD in using RFID to track material:

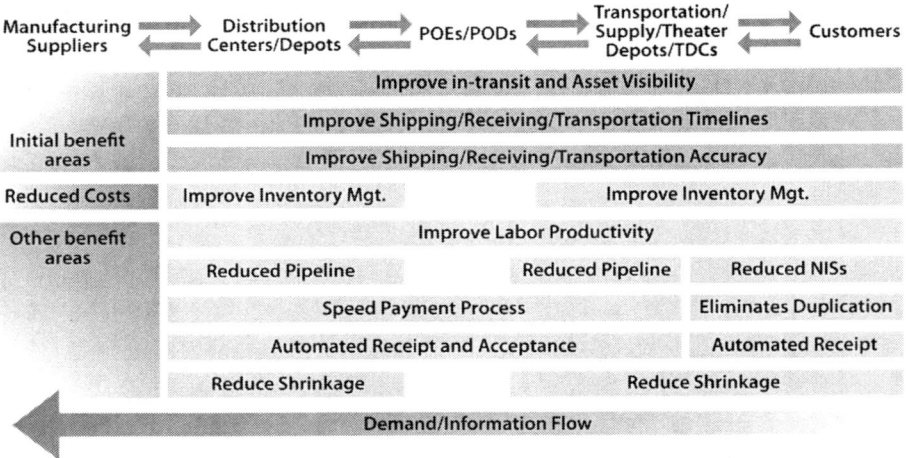

Given this emphasis on using RFID to actively monitor the movement of material, its not surprising that the DoD has established mandates to assure that its suppliers implement RFID tagging of packaging as quickly as possible—and once this tagging is in place, that the RFID information be inserted into its AIT workflow system by the vendor.

Accordingly, its RFID policy mandates that all vendors who are contractually obligated to affix passive RFID tags to material must also initiate an Advance Shipping Notice (ASN). This is the initial action that enables the DoD to "effectively utilize RFID events to generate transactions of record" at all levels throughout its logistics systems.

Fortunately for suppliers, this ASN is not a new process or transaction, but it is a variant of the existing Material Inspection Receiving Report (MIRR) transaction already being submitted—but with additional RFID data elements added. During the ASN transaction the sender relates the passive RFID tag ID (at various levels of detail) to the actual contents and configuration of a shipment.

Some details to keep in mind: vendors must provide the ID of every RFID tag in a shipment within the ASN, and they must represent this passive tag ID in a hexadecimal format. Typically, the hexadecimal format is the format used by passive RFID software in printers and readers, so the software should do the binary to hexadecimal translation process automatically. For more detailed information, vendors can consult the ***DoD Supplier Guide.***

Similar to the reporting of UID data to the UID Registry discussed in prior chapters, suppliers are required to submit the ASN through the Wide Area Work Flow. (WAWF). Once again, for flexibility the DoD allows a vendor to use one of the three methods available within the WAWF:

User Defined Format (UDF) via the vendor's File Transfer Protocol (FTP)
Interactive web-based site
Electronic Document Interchange (EDI)

Whichever method is chosen, the vendor will be completing five transaction sets:

    Receiving Report

    Receiving Report for Pack Update

    Combo Invoice & Receiving Report

    856 WAWF 4010 Detail Receiving Report

    856 Pack Update WAWF 4010 Detail Receiving Report for Pack Update.

These transaction sets have been modified by the appropriate DoD agencies to ensure the transactions can be used to list the contents for each piece of a shipment of goods as well as additional information relating to the shipment such as:

    Order information
    Product description:
    item count in the shipment piece and
    item UID information,
    Physical characteristics
    Type of packaging:
    container nesting levels within the shipment
    Marking information:
    shipment piece number
    RFID tracking number
    Carrier information
    Configuration of goods within the transportation equipment

One final note for vendors just getting started, you can access *WAWF guides and training materials* Even if you have not yet registered (see prior issues for registration instructions), you can still access the guides and training material by the following steps:

- Click on Logon to WAWF (Registered user only) **Note: Click this hyperlink even if you are not a registered user.
- At the WAWF—Logon page, enter the following account information: Either User ID = Vendor11 and the Password = Vendor1$ (Case sensitive) or User ID = Vendor22 and the Password = Vendor2$ (Case Sensitive).
- Click the 'Submit' button
- Once logged in, users should click on the link entitled: "FTP/EDI Guides & Other Supporting Documents" located in the menu bar on the left side of the Web page.

Chapter 9

UID + RFID + AIT= Win+Win

New Math leads to faster payment$ to suppliers.

Practically Speaking

Last chapter we covered RFID in general and its use for unique identification. Let's return to the application of this technology with a more extensive investigation of Passive RFID and how it relates to the UID requirements.

Aggressively Passive Activity

In 2007 UID Journal reported that passive RFID infrastructure implementation has been completed at all DLA distribution centers within the Continental US. Our discussion is centered on the use of passive RFID technology and how it is related to the automated transfer of item unique identification IUID data.

Passive tags have no internal power source and rely on RF energy transmitted from the reader to 'excite' their circuitry (ooh, love saying that). This energy is used to generate a brief digital response back to the reader. Because passive tags contain no battery their lifetime' is virtually infinite.

Though the cost of passive tags can be kept to a minimum, they may require stronger RF signals from the reader, potentially increasing its cost. More importantly, the signal strength returned from a passive tag is constrained to very low levels, so its effective read range is in terms of inches. However, unlike optical scanners, line-of-sight is not necessary to transfer the data. Read ranges can be addressed by building larger, more powerful readers but power consumption and costs increase exponentially. (Not to mention the eventual involvement of the FCC.).

Returning now to practical application, let's delve first into the use of passive RFID.

In conjunction with the more firmly established optical scanning technology, passive RFID is truly taking hold as an adjunct. In fact, both are required within DoD mandates MIL-STD-129, and MIL-STD-130-M. Which is not surprising, given that optical scanning is focused on identification of the actual, physical part; whereas RFID is attuned to gathering information associated with the packages containing parts with a 2-D data matrix. So, it's not like they are fighting over the same turf. And these two technologies work together in complementary fashion. This relationship is physically exemplified and readily visible in the latest generation of combination scanners, in a variety of hand-held or fixed formats, at the recent EPC show.

Most obviously, this enables the association and confirmation of packaging information with the identification of the specific part inside the packaging. Accordingly this operation is done upon the initial receipt of material and with the newer scanners can be accomplished simultaneously. Moreover, when combined with the growing capabilities of the integrated system software (*aka* middleware) and end-user applications, all sorts of contract information can now be automatically uploaded by secure FTP to WAWF and UID registry, by manufacturers/agencies/depots who supply the DoD.

Besides the immediate benefits of improving their processes, the integration and expansion of these identification technologies, helps defray costs that otherwise would probably be viewed by suppliers as being solely associated with meeting DOD requirements. Suppliers are encouraged to keep track of costs and assign them to the applicable accounting period. The cost associated with IUID is maybe allowable in old and new contracts. Check DoD Contract Pricing/Cost Accounting guidelines online (Source UID PMO James Clark PMP, 'IUID 101').

As mentioned earlier, one of the functions that RFID performs quite well is the reading of multiple tags. So it's ideal for answering the question as to how many cases are contained on a given pallet. This can certainly be accomplished by many of the latest model combination optical/RFID, handheld/fixed, and scanner/readers. And the fact that some are 'tethered' to a local power supply provides added mobility options for incoming and outgoing scanning stations.

Once you cut loose the bounds for a RFID reader by giving it a mobile power supply, interesting applications emerge. First of all, you can install the reader onto the cross-brace of the movable fork on forklift and read all the packages on a pallet while transporting it. One of the 'problems' with RFID technology is that, in a sense, it reads multiple tags *too* well—meaning that the reader is prone to also picking up the passive tags associated with another pallet. To address this limitation, 'finesse' is built into the mobile reader, in this case proximity and accelerometer sensors, which enable the embedded firmware to exclude extraneous tags.

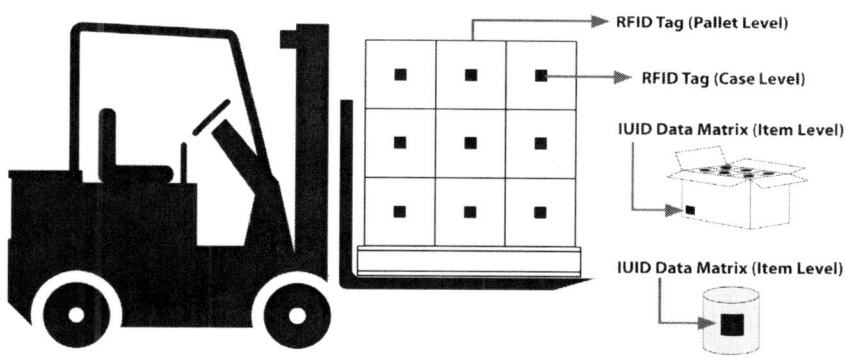

Implementation Plan: Level of Packaging

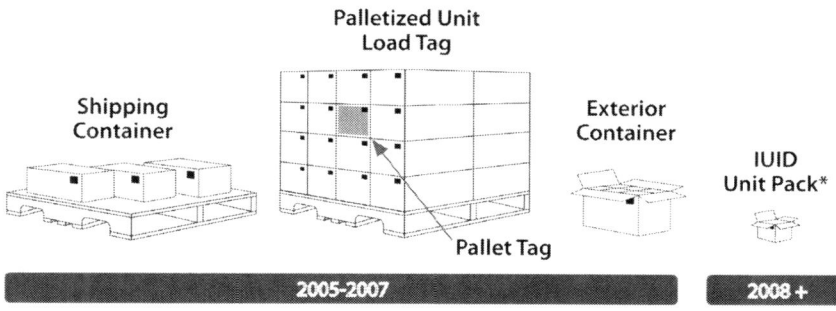

From a system viewpoint, such advancements in RFID application are important because they allow RFID technology to be utilized to automate DLA receiving and deployment of supplies. The more extensive this infrastructure is deployed and utilized the greater the efficiency and ROI.

Supplier Benefits:
- Improve planning
- Produce faster demand responses
- Reduce the Bull Whip Effect
- Streamline Business Processes
- Improve efficiency in the recall of defective items
- Increase your ability to ensure that your product(s) remain stocked on DoD's shelves
- Receive faster payments for your supplied goods

DoD Benefits:
- Improve inventory management
- Improve labor productivity
- Eliminate duplicate orders
- Replace manual procedures
- Automate receipt and acceptance
- Improve inventory and shipment visibility and management
- Reduce shrinkage
- Enhance business processes within the DoD
- Improve asset tracking

Initial reports indicate that DoD suppliers implementing compliance solutions get paid faster!

This initial benefit is far exceeded by the increased efficiency, associated with automated data capture, scanning and reporting, as well as 50-60% fewer items held due to Q/A issues (Source: Ford Powertrain 9/06).

Additional reference sites:

http://www.acq.osd.mil/log/rfid/index.htm

http://www.acq.osd.mil/log/rfid/DoD_Suppliers%27_Passive_RFID_Information_Guide_v8.0.pdf

http://www.acq.osd.mil/log/rfid/MIL-STD-129P-chg3-29Oct04%20(2).pdf

Chapter 10

Time Marches On: A Million Parts Later

From Zero to One Million

Sitting in the packed room for the Welcoming Address and Status Update at the Feb 07 UID Forum you could detect the collective feelings of both *acceptance* and *accomplishment* by the gathered participants.

'Acceptance', because as keynote speaker LeAntha Sumpter, Program Manager for Unique Identification, Office of the Under Secretary of Defense for AT&L pointed out, "Whatever your assessment of what's been attained through the unique identification initiative during the past four years, the one thing that everyone has come to appreciate is that it's not going away."

Which of course, garnered the expected wave of chuckling from the assembled audience. What you could not tell is the exact source of these releases. Did it come from pent-up frustrations? Wizened knowledge garnered from tough implementation challenges? Maybe just the overall burden inherent in any long-range effort aimed at changing the way we do business day-to-day.

No doubt, each of the 550 participants (not an official tally) at the Forum, could list their own *unique* reason to laugh, sigh or cry at their personal involvement in what Sumpter, assuming her DoD vestments, deftly reminded everyone is an "operational imperative"

Shades of Immanuel Kant! If I remember correctly from my Philosophy 101 class, then this puts Unique Identification right up there with concepts like

Space and Time. Which explains a lot. I don't know about you, but I'm still struggling with this whole Time thing—especially the implications regarding punctuality and mortality. Certainly explains the hint of nervousness detected in many of the chuckles—no getting away from ol' UID.

But like any good speaker, Sumpter quickly shifted the focus to more positive aspects—like what's been Accomplished. To bolster everyone's spirit, she displayed for the audience the following slide:

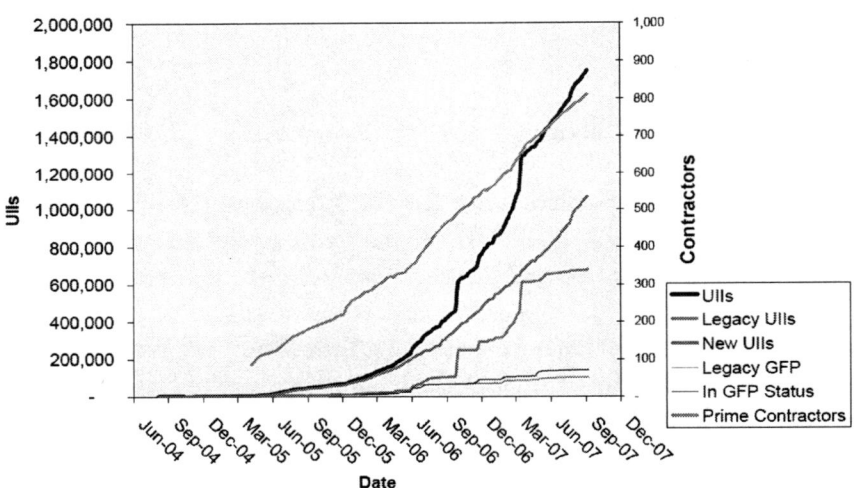

As of 2007 over 2 million items had been entered into IUID Registry

As of February 2009 over 6 million items had been entered into IUID Registry

First she directed attention to the steady additions to the Registry over the past years, the increasing slope of the diagram's curve over the last few years and how the number of items in the registry now exceeds 920,000, then confided to the group—with a slight note of disappointment—that she had hoped to be able to announce that the one million mark had been surpassed.

But take heart fellow campers. Maybe she couldn't proclaim meeting that milestone at that very moment, but in an exclusive journalistic coup (what we in the profession call a 'scoop') the *UID Journal* is proud to announce that by day two of the conference, according to our highly-placed source, Rob Leibrandt, Deputy Manager for UID, Office of the Under Secretary of Defense for AT&L, the **One Million Mark was Exceeded!**

In November 2007, The UID PMO announced that the IUID Registry had exceeded Two Million parts! Thus in approximately seven months they matched what had previously taken 4 years to accomplish. As infrastructure continues to proceed, we expect this growth to continue to increase.

Congratulations all around. Don't forget you read it here first.

Getting back to speaker Sumpter at her podium, she posed the pertinent question, "After four years of effort, what have we learned?" But before addressing her own question in detail, she paused for a moment of reflection and summation. "What we've learned in general is to not deal with each item in a 'special' way, but to handle all items the same way, using similar processes, while relying on the 'uniqueness' of the item identifier to handle it correctly."

More specifically, we've learned that many problems remain in need of fixing. Sumpter identified the following key challenges that must still be tackled:

- Item identification must be unique
- Machine-readable language for AIT devices must be interoperable
- Item valuation and property transfer processes must be consistent
- Maintenance data must be exchanged in a common format
- Standards are useful; but data keys remain critical
- Business processes must be Integrated across functional lines

Despite the encouraging numbers, Sumpter reminded everyone that there are still volumes of parts out there that need to be marked—and those parts that are marked now need to be tracked. "One thing that continues to amaze me is the number items that are not marked yet still tracked. In too many cases we're still relying on the old stubby pencil."

Marking leads to tracking, and tracking brings up the need for the expanded use of Automatic Information Technology (AIT) throughout the DoD supply system.

Then, once this information is in the system it has to be transferred more efficiently. Ultimately, this information must flow "end-to-end" through the broader application of AIT. "From creation of the part by the manufacturer throughout its total life cycle."

"More and more, program managers are coming to realize the huge people cost associated with poor data transfer." She then cited the continuing—and somewhat ironic—example that people in charge of disposing of an item typically have to scrounge around to correctly identify the part just prior to removing it from the system.

To address these fundamental issues, she emphasized that the current acquisition, finance, and logistics business rules in place with have to be "re-engineered" to assure that items are both uniquely identified and appropriately tracked. Leading this effort falls to the Program Managers, which she pointed out is why so many of the sessions at the UID Forum focus on the role of the Program Manager and their work with the engineering community, contractors and depots. Once this new system is attained it will then produce the all important management information. "Data, it's all about data."

Looking beyond the struggle to get each of the fundamentals right, Sumpter called attention to what she sees as the greatest reward that will follow—the "gold mine" of information available. As the earliest example of what can be gained from attaining a "data rich environment", she noted the results already achieved from the use of the maintenance information, especially regarding warranties, "Truly, one of our greatest success stories."

Accordingly, when she presented the audience with a glimpse of "what is on the horizon," there was heavy emphasis on the exchange of maintenance data. This will include the demonstration of the Product Life Cycle Support (PLCS) Data Exchange Set (DEX) for aviation maintenance along with the mapping of 13 key processes for the maintenance community. Supporting this effort will be the adoption of ISO 10303 and Application Protocol 239, defining PLCS and enabling enterprise data integration; while leveraging the use of Warranty information will continue through the flagging of parts with express warrantees and the tracking of warranted parts.

Sumpter reminded everyone present, regarding the forthcoming adoption of the proposed DFAR clause mandating use of the Wide Area Work Flow WAWF as the only DoD payment system by mid-2007. No chuckles were heard following that sobering moment. But Sumpter concluded her presentation by encouraging everyone to savor what successes have been achieved so far the initiative and from their hard word—and listed the following:

- Maintenance data partnership between Army Material Command and Naval Air Systems Command
- Air Force Pathfinder Projects
- Integration by Defense Medical Logistics System
- Data Integration by TACOM
- Tinker, Corpus Christi, Albany Depot Progress
- Marine Sense and Respond pilot
- Integration with Loss Damaged and Destroyed and Plant Clearance Systems (DCMA)

Chapter 11

UID PMO Policy Issues/Information

'A Revolution in Total Asset/Item Visibility'

UID: It all starts with a Part Mark

The UID PMO Keynote Speakers at the UID Forum San Diego February 21, 2007, included Rob Leibrandt, Deputy Program Manager for UID, US Department of Defense. Mr. Leibrandt gave the following overview of UID:

> 'In today's world, many items we buy have identification that indicate the manufacturer or distributor of the product and the product type UPC (Enterprise ID). Additionally, some items also have a serial number that differentiates one item from another identical item. It is the combination of Enterprise Identifier with a Serial ID that a globally unique item identifier is created-to differentiate every item from other items. Thus, each item has its own globally unique identifier (UID).'

Making the Item using the ECC200 Data Matrix Symbol (renders the information Machine Readable, or MRI) with data elements that follow supplier's business processes has considerable benefit:

Gathering and delivering the required data, Marking the Item using the ECC200 Data Matrix Symbol with data elements that follow supplier's business processes has considerable benefit:

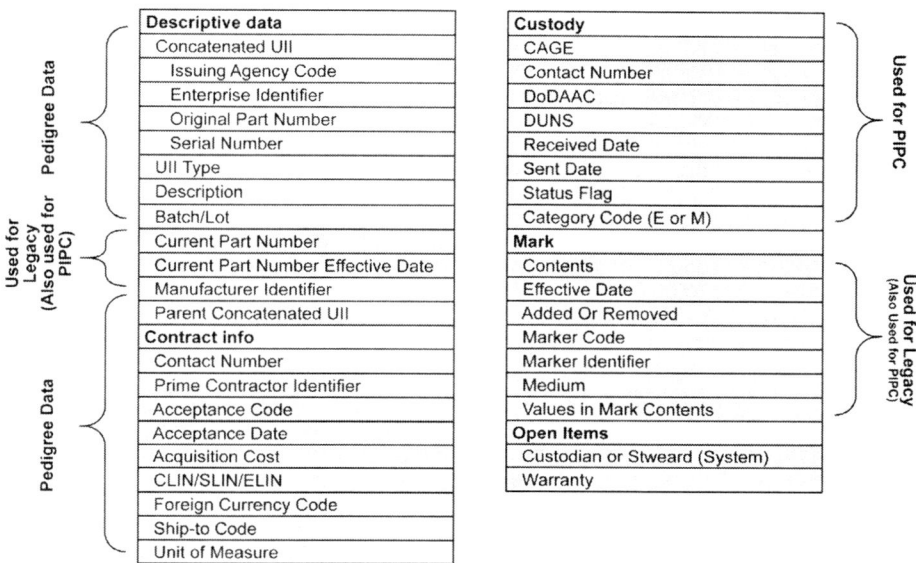

MRI Significantly reduces risk of quality failure associated with misidentification.

Eliminates legibility issues.

An enabler to a paperless system.

Improves speed and accuracy of data transfer.

No data transcription errors.

Internationally recognized.

Has the ABILITY to:

Improve parts traceability

Reduce internal processing procedures

Capture accurate 'As Built' data

Check 'Should Build' data

Reduce Replenishment costs

Generate electronic logbooks.

The Focus Areas For the UID Forum included:

Marking the item-using the ECC200 Data Matrix Symbol

Gathering and delivering the required data

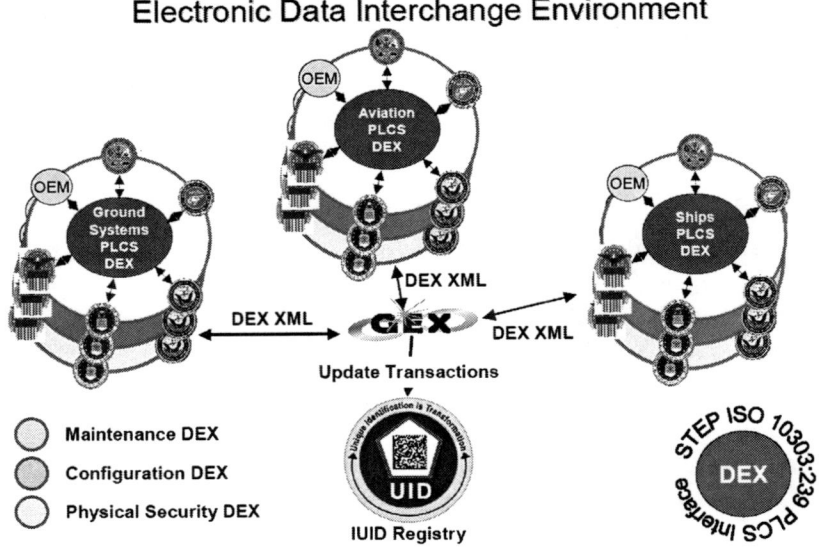

Exchanging data with other stakeholders

Leveraging the AIT and the data to improve supplier's operations.

When You Have the Capability to:

Identify items individually and uniquely (IUID)

Exchange data in a common format (ISO 10303—STEP/PLCS)

Use the identification and exchange to "Track" an Item

Then You can realize Process Improvement Opportunities:

Improved Data Quality through Automated Identification and Data Capture (AIDC)

Improved Inventory Control

Better Warranty Management

More Rapid Fraudulent Parts Detection

Identification of "Bad Actor" parts

Efficiencies from Corporate Serialization Standardization

Improved Through-Life-Visibility

Increased Reliability and Availability

Implementation of a Predictive Maintenance model versus Use-to-Failure model

Chapter 12

UID Implementation & Contract Retention

While Monterey, California may be most known for its breathtaking scenery and world-class golf, when it comes to taking good shots along the Central Coast, there is one place, which is worthy of equal praise—**Laser Devices, Inc.**

This international manufacturer is a leader at the forefront in the design, implementation and production of a wide range of aiming lasers, illuminators and tactical lights-and has fulfilled UID requirements.

Laser Devices, Inc. (LDI) recently won a significant portion of the U.S. Department of Defense Multi-Functional Aiming Light Contract #W91CRB-05-D-0029. This contract was awarded for the manufacture of the Company's DBAL-A2 (Dual Beam Aiming Laser—Advanced). The DBAL-A2 was selected for acquisition by the U.S. military based on its ability to supply visible and infrared aiming lasers along with an infrared illuminator that our troops can use for pointing out targets and for accurately aiming and illuminating targets.

While LDI's initial contract, awarded in September 2005, did not specify a UID requirement, it became apparent in late 2005 that the DBAL-A2 would require UID compliance to meet the U.S. Department of Defense (DoD) mandate to fully implement UID on new equipment acquisitions.

LDI and the DoD agreed upon a contract modification in early 2006, requiring UID implementation based on MIL-STD-130M. LDI's Logistics

Manager Alan Movson, was assigned responsibility to develop and execute an action plan so that LDI would be able to meet its UID requirements.

The DBAL-A2 housing is machined out of aircraft aluminum to meet MIL-SPEC-810F for reliability under the most rigorous environmental conditions. In addition, the DBAL-A2 is waterproof to depths of 60 feet and can operate in the most extreme climatic conditions from the most arid deserts and snow-covered plains to the highest mountaintops. External azimuth and elevation mechanisms allow the aiming lasers to be accurately aligned for use on a variety of military weapons.

One of the challenges was to mark the product in a way that the 2D bar code would remain legible for the life of the product. It was vital to evaluate the impact and effectiveness of the various options for UID Direct Part Marking such as laser, etching, dot peen and labels. Given the continuous and daily physical strain placed on the product during training and combat missions it was decided that etching the UID into the surface of the DBAL-A2 housing was the preferred option. LDI's subcontractor was responsible for anodization and agreed to etch the units and also verify that the UID labels conformed to MIL-STD-130M.

Laser Devices soon recognized that having come to a decision on the method of etching, the hard work was just beginning. As a small business that places a premium on efficiency it became important to find a single integrated solution to efficiently capture UID information for the following:

- Use in the government system for electronically processing invoices using the DD-250 form Wide Area Work Flow (WAWF)

- To be able to use the WAWF to transmit the product data encoded in the data matrix directly into the UID registry

- To use this information while generating the unit packaging labels required by Mil—STD-129P

After substantial research, LDI selected two vendors in the industry to provide an integrated and cost effective solution that would have minimal impact on overall productivity. LDI has quickly and successfully implemented the UID solution as required by the contract. LDI believes that the integrated solution will be helpful for other small manufacturers who are looking to implement UID in the future.

Chapter 13

Effects of UID Implementation on Manufacturing WIP

"We've always seen this as an opportunity to reduce quality failures associated with identification. A '5' and an 'S' will never be confused again because it is being read by a machine." reports Nat Rushard, Rolls-Royce Team Leader (UID Journal July 06).

According to John Piatek, VP Freedom Technologies, 'The goal is to remove the human decision from the manufacturing processes. According to Ford Powertrain's Warren Schwanky, 'The goal is to use UID Serialization early on in the manufacturing process to create a complete 'birth history record' for life-cycle of the individual parts'.

Manufacturing Processes affected by the interface of UID Data include: Receiving, Assembly Floor, and Quality Analysis (QA).

One result is greater shop floor cell traceability, production visibility and ultimately thru-put. Another result is lot traceability, individual operator accountability, cycle count visibility of inventory, process verification, containment of parts as they pass through line at sub-assembly and final assembly, child-parent part verification via interface to all shop floor data devices including PLC's, Vision Systems, and sensors.

The UID Part Mark is written onto the part once, and read (scanned) many times thereafter. Therefore, the UID partmark, and its location must be documented for maximum effectiveness,' according to Ford Motor Company Powertrain Supervisor, Warren Schwanky ('Beyond Compliance' Webinar, co-sponsored by UID Journal, Aug.23rd, 2006). He uses the example where a UID Part Mark on engines was re-located so it would be accessible to line-of—sight scanners when manufactured products were transferred to inventory racks.

Equipment needed by manufacturers for integration into manufacturing process-varies according to the what materials comprise the products to be marked—include a combination of marking equipment including: label printers-inkjet or thermal; or lasers, dot-peen markers, chemical etch; scanners-fixed (on the assembly line),hand-held, batch, hand-held, wireless or cellular, high performance, auto-discriminating, variable focus, combo-RFID, UID, Barcode, 2-D—and software database to report UID and contract data to Wide Area Workflow or UID Registry (or both), and store validation records, reports and gradients to prove the UID met specification. Some additional software benefits include: automatic association: that parts are correctly manufactured, that parts are correctly marked, and verification that parts are not 'on-hold' status.

One of the common requirements for manufacturer's engineering departments is to update documentation to include changes. Implementation of UID involves documentation changes that companies might find challenging:

UID is extremely valuable on the manufacturing floor from start to finish. The key is one identifier (UID), all through the process. Companies will take all this data and make higher quality parts, better, faster and more cost effective' according to RF Technologies.

Although Industry experts indicate the part should receive a UID part mark at the earliest possible point in the manufacturing process, According to John Piatek VP Freedom Technologies 'UID Validation should occur in the final phases of manufacturing process and immediately prior to the part being shipped when data is sent to the WAWF'.

In addition to the quality gains, however, a premier engine manufactures says that time savings have been a subsidiary benefit.

"Take for example the traditional method of data capture for fan blades. Fan blades are very finely balanced. You can spin the whole of the fan with your index finger. To do that, we have to collect data from the manufacturer to balance each fan accordingly. There are typically 26 fan blades and it used to take 90 minutes to gather the information manually. Now it takes around 3 minutes. So there is definitely a cycle time benefit." In addition to the speed and accuracy of the system, it is also an enabler to a paperless system. "Once the data is gathered, it then goes to a certification group, who will create the as-build conditions. They have to re-type and re-format the data, but with machine-readable data, it can go straight into the system. With direct part marking, we can capture both the as is-built condition and check it against the should be-built condition. Have I put the right parts in the right engine? Direct part marking enables us to answer this question with greater accuracy. There have been many secondary benefits, but this—the improvement in accuracy and quality—has been the primary aim and payoff."

Additional benefits resulting from the automated data capture of the manufacturing Process Q/A database allows for trend analysis by manufacturing and linking of quality tests (including Pass/Fail) to item serial numbers. Potential Product recalls can be pinpointed to specific part, model, manufacturing cell location and specific customer.

The ability to focus on the entire life cycle of an item, beginning with the 'as-built' history of that item during the manufacturing process, and consisting of a complete historical record of all activity including: Stress testing, analysis, maintenance. 'The results are a Dramatic Cost Reduction in number of quarantined parts and man hours', according to Ford Motor's Schwankey.

Rob Leibrandt, Office of the Undersecretary for Defense AT &L cited 'ADC (automatic data capture) capability offers a huge advantage, for example, Rolls-Royce found 3-4% of all direct labor costs associated with manufacture of aerospace engines were directly attributed to manual entry of part item data collection.'

This substantiates the assertion that a UID Compliant Implementation would yield as an additional benefit, a return on investment (ROI). Although compliance may be viewed as a cost, the use of the technology can yield numerous actual benefits. Although a company may initially find planning a UID Implementation somewhat challenging from the perspective of where to locate the equipment, the additional efficiencies affect the entire product manufacturing process. In the final analysis, the end results more than justify the means!

Chapter 14

Validation, Verification and Grading

What is it, Who needs it and Why?

V**alidation** is used to confirm that a code's data is encoded in the correct format structure and syntax to conform to the accepted Standard (MIL-STD-130M).

Verification is a physical measurement of the quality of the mark. Verification is the process of inspecting and documenting whether marks conform to the accepted standard (MIL-STD-130M). Why is it required? Also marks must conform to MIL STD 130M specific requirements for mark quality. The contracting officer and the contracting language in DFARS 211.274-2.

*What about **costs** for the implementation?*

Count your costs/assign them to the applicable accounting period. The cost associated with IUID is allowable in old and new contracts. Check DoD Contract Pricing/Cost Accounting guidelines online (Source UID PMO James Clark 'IUID 101').

Verification Validation and Grading—

We asked Matt Van Bogart, Product Manager, *Microscan* to explain: "The most successful way to ensure complete quality control of your production process is to directly mark products with a machine-readable code and enable tracking through its entire life cycle. This method is required on nearly all products supplied to the Department of Defense (DoD), which requires Unique Identification Data (IUID) to be encoded into a 2D Data Matrix symbol and marked directly on specific tangible items. Poor quality marks can lead to the rejection of your parts and components based solely on illegible IUID information. DoD suppliers rely on verification hardware, as manufactured by Microscan (i.e.,Microscan's Quadrus® Verifier), Cognex or Siemens, to analyze and grade direct part marks using the symbol quality standards identified by the DoD's MIL-STD-130M. ISO/IEC 15415 is an included standard containing several important quality criteria.

Symbol Contrast

Good contrast creates a strong signal, which makes it easier for the imager to differentiate between the light and dark elements of the symbol. This will make it easier for the imager to read the code as well as reduce the chance of noise interference. High contrast also increases the ability to read at longer distances.

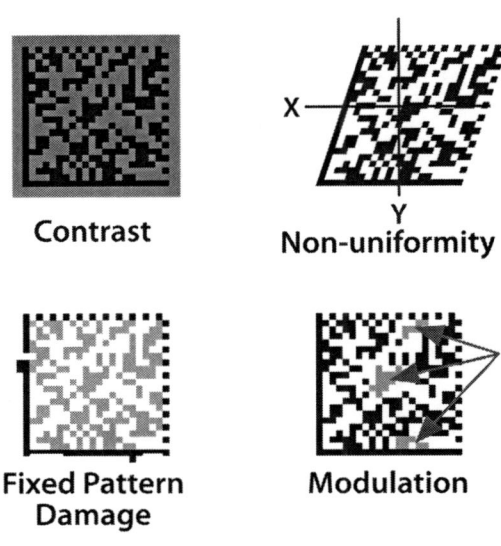

Contrast

Non-uniformity

Fixed Pattern Damage

Modulation

Unused Error Correction

Axial and Grid Non-Uniformity

Axial non-uniformity measures the consistency in spacing between the elements in the major axes, and grid non-uniformity measures deviation from a perfect grid. In order for a Data Matrix to be readable, it's important that the elements are consistent throughout the symbol, forming a perfect square. If the symbol isn't square, the distortion could be caused by slack in the marking system or from marking around a curved surface.

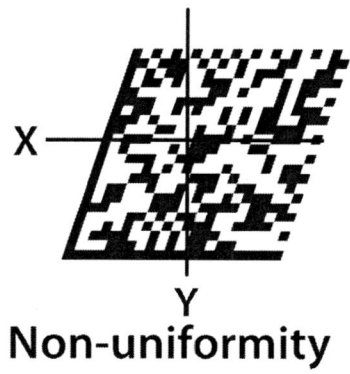

Non-uniformity

Modulation

Mark consistency is critical in creating a readable code. Modulation refers to the ability to discriminate between light and dark elements within a symbol. If the dark elements appear in varying shades, this can reduce a reader's ability to properly distinguish between elements and result in an unreadable mark.

Modulation

Fixed Pattern Damage

Data Matrix symbols contain fixed patterns, used for locating, orienting and mapping the symbol. Distorted or missing patterns will make any symbol difficult or impossible to read.

Fixed Pattern Damage

Unused Error Correction

Be sure to select the ECC 200 type of Data Matrix, which incorporates the robust Reed Solomon method of error correction and ensures the maximum data security for your symbol. If your symbol becomes damaged in some way—scratched, missing corner, ink blotches—the symbol may still be readable. And since most products in the industry have standardized on ECC 200, you'll have more hardware options available.

Unused Error Correction

Meeting the quality requirements of the ISO/IEC 15415 standard is just one of the requirements for direct marked parts supplied to the DoD. Understanding quality expectations and the available verification hardware options will help increase your accepted parts and positively impact your bottom line.

Chapter 15

Direct Part Marking 101

An Overview from an Expert

We went back to a familiar source, Matt Van Bogart, Product Manager, Microscan, for the basics on Direct Part Marking:

Electro-Chemical

This marking process uses a low voltage electrical current to pass through a stencil to the part's surface. In order for this method to work, the part must have a conductive metal surface. This method will not work for anodized, powder-coated or non-conductive coatings. Unlike other permanent marking methods, electro-chemical etching does not weaken or distort metal parts because the molecular structure of the part is not altered beyond the depth of the mark. As a result, very thin-walled parts and those with fine surface finishes can be safely marked without damage. Since electro-chemical etching is a more involved process than other methods, it is not suited for highly automated applications and is commonly used for low volume product runs.

Ink Jet

This type of marking uses small, circular dots that are sprayed directly onto the surface of the part. Ink jet typically produces high contrast marks, depending on the substrate and the ink color. Although permanent inks do exist, ink jet is not considered by some industry standards as a permanent marking method. Take care to ensure that you select the most appropriate ink for your substrate. Disadvantages include routine maintenance to prevent the jets from clogging, and the additional cost of consumables.

Laser Etch

This marking type uses lasers to etch the symbol directly into the surface of the part. In addition to producing a clean, high-resolution mark on a variety of substrates ranging from metal to plastics to glass, laser etching is also well suited for automated environments requiring high volumes. Since the top layer of the part's substrate is removed during the etching-process, sometimes the minimal residue that results may not be suited for some clean-room applications. The type of laser (Yag, CO2, YVO4) must be matched to the application and will affect price considerably. While laser-etching equipment has a higher entry cost than many marking methods, there is no additional cost of consumables and maintenance is minimal.

Dot Peen

Dot Peen is a percussive marking method, using changes in depth to create the contrast between the light and dark elements of the symbol. Dot peen is recommended for applications where the symbol must last the entire life cycle of the part. In the aerospace and automotive industries, this can be several years. Suitable substrates for dot peen marking must have some hardness so material memory does not return the surface to its original condition.

Additional Marking Methods

Methods available include metal stamp, engraving, electrical arc pencil, embossing, cast or forged (bumpy bar codes), molded, rubber stamp stencil, and decalcomania. While not as common as the four methods previously discussed, these are all viable marking methods for creating direct part marks and are useful for specific applications.

Marking Method Selection

Before you start evaluating marking methods, first determine if an industry standard applies to your application. Industry standards will often times specify specific marking methods to be used, in addition to symbol specifications.

Marking Method Grid

Marking Method	Description	Advantages & Disadvantages
Ink Jet on substrate 	Contrast levels vary widely, round element shape **Application:** - Post-packaging - Warehousing - Automotive	**Advantage:** - Low-entry cost - High speed - Easy to read if contrast is good **Disadvantage:** - Not considered permanent by some industry standards - Dot registration can vary - Higher cost consumables - Mark quality dependant on surface cleanliness - Difficult to read if contrast poor
Pre-printed packaging 	Typically high contrast, square element shape **Application:** - Product labeling - Product packaging - Document processing	**Advantage:** - Economical - High speed - Good contrast - Easy to read **Disadvantage:** - Less flexibility
Thermal transfer label stock 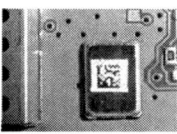	High contrast, typically black on white label stock Square element shape **Application:** - Product labeling - Packaging - WIP tracking, various industries	**Advantage:** - High contrast - Low-entry cost - Easy to read **Disadvantage:** - Not permanent - Higher cost: consumables
Subsurface anodised printed label by digital technology 	High to Medium contrast **Application:** - MoD, DOD, AWE, Aerospace, Medical device, asset control and public Services.	**Advantage:** - Rugged solution, Permanent, Extremely durable, High quality mark, No surface damage, No consumables, performs in normal right up to the most hazardous environments. Can be printed on a wide range of thicknesses e.g. foils up to 3mm plates. **Disadvantage:** - Can only be printed into Aluminium.

Marking Method	Description	Advantages & Disadvantages
Laser etch on silk screen 	High contrast, square & round element shape **Application:** - Electronics	**Advantage:** - Good contrast - No consumables - Permanent **Disadvantage:** - Displaces surface - Process creates debris
Ink jet on plastic 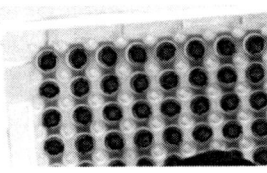	High or low contrast, round element shape **Application:** - Bio-science - Pharmaceuticals - Packaging	**Advantage:** - Limited damage to surface **Disadvantage:** - Higher cost consumables - Not permanent - Bleeding can affect mark quality
Laser etch on rubber 	Very low contrast, square or round element shape **Application:** - Automotive	**Advantage:** - Permanent - No consumables **Disadvantage:** - Process creates debris - Affects surface of substrate
Chem etch on metal 	Typically medium contrast, square element shape **Application:** - Electronics - Semiconductor - DOD - Aerospace - Medical device	**Advantage:** - Permanent - High quality mark - No debris from process **Disadvantage:** - Potentially toxic material bi-product - Low-volume use only - Potentially complex process

For your reference, the Marking Method Reference Grid provides examples of the more common DPMI marking methods in use today. If a specific marking method is not specified or recommended for your application, then use the following guidelines to help you determine the specific needs of your application. Important factors to consider are permanency, material composition of the part, manufacturing process, cost of the part, and production volume.

Material Composition

Be sure to select a method that will have the least amount of impact to the composition of your part. Experiment with different depths. Sometimes this can make the difference in success or failure of a method. If your part has a thin wall, you may want to avoid dot peen as it may permanently damage your part. If your part is powder-coated, electrochemical etch will not work.

Permanency

Determine how long the symbol needs to last. This is often dictated by what is encoded in the symbol and what it is used for. If the symbol is used strictly for in-house work-in-progress tracking, than it need only survive the manufacturing process. For this type of application, ink jet may suffice. If it will be read repeatedly throughout the supply chain or for the entire life cycle of the part, than a more permanent method such as dot peen may be necessary. Keep in mind life cycle is defined differently, depending on the industry and the part.

Production

What is the production environment? Is it a highly automated facility? Will most of the codes be read by manual presentation? Some marking methods are better suited for high-volume production, such as ink jet and laser etching. Electro-chemical etching is better suited for low-volume or semi-automated environments. Consider how the marking method may affect output capabilities and potentially affect the cost of manufacturing.

Manufacturing Process

Determine where in the manufacturing process your part will be marked. How will the various steps affect the integrity of the symbol and its readability? For example, a manufacture of medical devices laser etched a 2D symbol on the titanium case of a device. After the device went through the wet-blasting sterilization process, the appearance and contrast levels of the 2D codes changed. These types of changes can be easily accommodated by smart cameras or imagers designed for reading a variety of code types. However, if your part is painted after it is laser etched, the laser-etched mark may not have enough depth to still be readable.

Chapter 16

From Offense to Defence: RFID at the DoD

The purpose of RFID technology has changed a lot since its inception. In its earliest days, RFID was used as a tool of espionage during the Cold War. The Soviet government used it as a covert listening device that retransmitted radio waves with audio information.

Today, RFID is known less for its eavesdropping ability, and more as a way to add efficiency to some of the world's largest and most sophisticated supply chains. This strength is a key reason why, in 2004, the United States Department of Defense (DoD) signed a memorandum outlining a policy for its use. The DoD looked to the technology to play a key role in its broad Automatic Identification Technology (AIT) strategy, which is designed to help the organization more quickly and easily track and record shipment data, as well as the actual shipments themselves.

The RFID policy addresses both active tags, which are battery-powered, and passive tags, which "wake up" and generate a response in the presence of an RFID reader. The policy directed the adoption of high-data capacity active RFID throughout the DoD's operations, and mandated that suppliers include passive tagging at the case and pallet level for specified products by January 2005.

As a public sector entity, the DoD is accountable to taxpayers for its expenses. One of the main reasons behind the RFID mandate was to reduce the costs associated with inventory shrinkage, the tracking of missing items, and the shipping costs incurred in the event of incorrect or duplicate orders.

As well, it is important for the DoD to maintain constant control over its assets, and to have visibility into the status of its inventory. By doing so it can prioritize shipments, so that material destined for forces in war zones are able to get what they need when they need it most.

Finding a Solution

When implementing RFID, it's important to have key initiatives in place beforehand in order to ensure a smooth transition. In the DoD's case, this meant preparing for the infrastructure changes that would have to be made to 19 distribution sites across the United States. Making those changes posed a challenge, given an aggressive implementation timeline.

The DoD started by putting in place the foundation required to make RFID work effectively. To do this, the DoD installed fixed portal readers at warehouse entryways and implemented software running on rugged handheld computing devices designed to withstand all types of conditions, whether at a depot or on a base. The portal readers, recording data from each tag, scan the RFID tags, located on all incoming and outgoing products. The data collected includes everything from what is in the box to where it is located in the warehouse. The RFID software, configured specifically for the DoD, allows operations staff, regardless of where they're located, to monitor and manage shipments throughout the DoD's facilities. It also enables them to quickly and easily maintains optimum functionality across the entire network of devices. This means the devices can instantly communicate with other readers and printers.

The implementation of new technology however, does not simply mean that challenges will magically sort themselves out. The successful results seen at the DoD required lots of planning and legwork. Effective project management was critical to the RFID project. Hands-on training and end-user education were additional keys to success. As part of this effort, regulations and specific instructions for suppliers were presented in the *DoD Suppliers' Passive RFID Information Guide*, which is posted online at *www.dodrfid.org*. Additionally, DoD conducted training for Procurement Technical Assistance Center (PTAC) representatives to enable them to assist small- and medium-sized businesses who need to be RFID-compliant.

Seeing the benefits

Today the DoD is enjoying the benefits of its current RFID technology, which has helped it to dramatically streamline its supply chain. As well, the DoD is able to better manage inventory and, ultimately, reduce operating costs. In addition, they continue to seek out and leverage other benefits from RFID in the future.

The DoD staff now work faster and more efficiently on the warehouse floor. The sending and receiving of receipts and acceptance documents have been automated, and duplicate orders have been reduced, saving the DoD valuable time and money. Another key benefit, and one that is particularly important to the DoD, is the ability to temporarily associate "conditional state" information about a specific item that is received or waiting to be shipped. This information helps workers identify an item and see a roadmap for where it is going. The staff knows right away whether an item is destined for transportation, supply management, maintenance, distribution or disposal. Sensitive or important materials are now easier to locate and monitor.

Suppliers are also enjoying the collaborative benefits of RFID. The ability to gather and store information has produced a faster response time for product shipments, while enabling them to maintain a fuller product stock and quickly recall defective items when necessary. Suppliers also enjoy speedier payments for their goods, which makes working with the DoD a smoother experience.

RFID has proven to be an effective and beneficial technology for the DoD. By streamlining its supply chain, the DoD is able to manage its inventory in a way that's best suited for optimum levels of national defense. But perhaps the biggest benefit of all is the ability for the DoD to get the proper material to personnel in combat zones, at the right time, and in a condition worthy of the important job they are doing. In this way, RFID not only helps support workflows and inventory, it also supports DoD most valuable asset—its people.

Chapter 17
Keeping Track of Other People's Property

All qualified government property, was required to be marked and reported to the IUID Registry by Sept. 30th, 2007.

Each year our Federal government supplies millions of dollars' worth of property to contractors for use in performing their contracts. The total accumulated amount for such Government Furnished Property (GFP) easily reaches into the billions of dollar, so it's not surprisingly, that the Federal Government feels obligated—and is extremely motivated—to account for the whereabouts for all this property. The stated requirement that all qualified GFP property, including government Property in the Possession of Contractors (PIPC), be assigned item unique identifiers and entered into the IUID Registry by September 30th, of 2007.

Such government property generally falls into one of the following categories (although all may not require IUID):

- Material
- Special tooling
- Test equipment
- Property designed for military operations
- Facilities (used for production, maintenance, research, development, or test purposes)

It should be noted that GFP can also take the form of data or any information required for contract performance. For the literal minded . . . Subpart

45.1—General 45.101 Definitions. "Government-furnished property," means property in the possession of, or directly acquired by, the Government and subsequently made available to the contractor.

Whatever its form the use and possession of GFP by a contractor brings with it specific obligations to account for this property. But the relationship between the government and a contractor is also a two-way street, because both parties retain certain rights concerning the GFP.

These rights, in turn, create the potential for a wide variety of contract disputes, such as:

- Who is responsible when late or defective GFP affects the contractor's performance?
- What preconditions must contractors meet to be entitled to recovery?
- Who is responsible for the loss, damage or destruction of GFP?
- What methods may be used to price the claim, once liability is established?

To expediently resolve these inevitable disputes, a system must be in place to correctly convey the status of this property and to the answer the typical questions such as:

- Where is it located?
- What is its physical condition?
- What's it current value?

As with any accounting scheme, the system requires that the physical assets are first identified and tagged; but the true effectiveness of the system comes with the ability to then track the material. In this regard, the requirement to tag and track GFP parallels the requirement already established by the DoD to tag and track property created by contractors—and to do it as efficiently as possible.

Accordingly, the policies in effect for handling GFP are extensions of the overall policy that the DoD has developed regarding items in its supply chain, including, most importantly, the mandate for use of Item Unique Identification (IUID): Initiated in 2003. The DoDs general UID policy has produced several key policy documents including: Final IUID DFARS (Defense Federal Acquisition Regulation Supplement) Rule Published Dec. 23rd 2004 and May 12th 2005 Policy for Unique Identification (UID) of Tangible Personal Property.

The salient point of these two policy documents is the stated requirement that all qualified property, including

> GFP be assigned item unique identifiers and entered in the IUID Registry by September 30, 2007.

How this relates specifically to GFP, is stated in the policy document GFP IUID Policy Memorandum dated May 12, 2005. The objective of this policy is to attain a 'paperless' process for dealing with GFP, so it contains the following points: Directs expansion of the IUID Registry for electronic GFP management Directs expansion of the Wide Area Work Flow (WAWF) to capture property transfers. Contains

Currently, the Proposed DFARS rule changes dated March 21, 2006, is undergoing revision to address issues identified in the comments received. Final approval resulted in publication of the final Rule covering the use of IUID for GFP. Tasked with taking all this forward is Lydia Dawson, Senior Procurement Analyst, UID Program Office, Office of the Under Secretary of Defense for AT&L, who points out that contractors should not wait to begin implementing improvements.

Either way, the UID program office is geared up to help contractors now. "Everyone here has already had their baptism by fire. During the past year, we've been through a complete cycle—the first spiral—of conversions. "One of our main goals is to help contractors migrate to the electronic reporting of GFP by utilizing the IUID Registry."

This requires the phasing out the of current, paper-based reporting because it does not provide "sufficient and complete" information regarding GFP." Moreover, adoption of electronic-reporting "will enable greater fiscal accountability from the managers of DoD property." In practical application, the UID Program office is currently working with contractors and program managers on initiating and improving the following key processes:

> Identification of GFP items that should be marked

> Assignment of globally Unique Item Identifiers (UII) for these items

Entry of the UIIs into the IUID Registry using the IUID Registry to report GFP results in elimination of DD Form 1662

[Use of the WAWF for reporting transfer of property has been detailed in a prior chapter "Reporting the Data"—ed. note]

In the past, contractors have used DD Form 1662 to fulfill the requirement for annual reporting of all GFP in their custody. This form came as a response to earlier congressional hearings that were sometimes critical of the DoD for being 'unaware' of the amount of GFP in the hands of its contractors.

Though an improvement, the form still has several deficiencies: Only provides summary-level data; Does not consider capitalization requirements or useful life; Does not produce information of the existence, completeness, or specific item valuation. This can result in double counting of reparables.

According to Dawson, "The main emphasis for our office during the past year has been on two programs in parallel, to assist prime contractors in correctly entering items into the Registry and to guide them through their transition away from DD Form 1662." Though the deadline was September 2007 (all qualified PIPC must be entered in the IUID Registry), it's important to keep in mind that the program for converting these processes has been underway for the past several years.

The elimination of using DD Form 1662 is a good example. Technically, the 'transition' that Dawson mentions was initiated as far back as May 11th 2005 in anticipation of the use of DD 1662 being 'eliminated' for new contracts and solicitations being starting in September 2006. This triggered the first round of conversions by participating contractors who begin using the IUID Registry instead, as a DFARS "approved substitute" for the paper form DD 1662. Helping these initial contractors convert is the 'first spiral' that the UID Program office endured and learned from.

Round 2 of this conversion process began October 2006 and the program office expects a sharp increase in the number of participating contractors. Indeed, things are moving along. After the first round of implementation ending September 2006, a total of 85 contractors had contributed approximately

69,000 items to the Registry. In the second round, the program office reports that the number of participating contractors has increased to more than 800 contractors by October 2006 and more than 100,000 GFP items have been entered into the Registry. Dawson is quick to remind remaining contractors that this "period of transition" will not last much longer and that the "time to act is now," because ultimately "the onus is on the contractor to comply."

So, how do you know it your one of these contractors that "must comply"? If you are a prime contractor, chances are, you know already. The Defense Contract Management Agency (DCMA) has already contacted the 'bulk' of contractors with GFP in their possession. Notices have gone out over the past few years, but mainly to prime contractors. So if you are a sub-contractor, you may want to contact your prime and pay close attention to the details of your contracts, because the other "method of communication" is through the wording of these contracts. This should help you avoid any ugly surprises.

Regarding the possible consequences for contractors who fail to "make the best of the transition period", Dawson reminds everyone that they are just forestalling the inevitable and losing precious time. "Electronic processes have been modified to track custody and stewardship. If there is a change in form, fit or function, the part number, date and cost of the change is to be tracked in the IUID Registry, as are lifecycle events such as disposals. The requirements for GFP are the same as for new items."

Besides, as Dawson points out, the requirement to implement these new processes is in line with the internally driven, better business practices that companies are adopting anyway to become more efficient and stay competitive in the marketplace. "As a contractor, the last thing you want to do is fall too far behind the curve, because when the final Rule is adopted, there will be no 'grace period'." So, the best strategy for a contractor is to get involved sooner than later.

Plus you get the benefit of being in on the ground floor, while time and patience are more accessible. "Testing interfaces are still being worked on. A number of contractors have expressed an interest and willingness to participate in this process. Additional contractors who are interested in doing this should direct their inquiries to the UID Help Desk." Phone them at (703) 848-7314 or email them at: *info@uniqueid.org*.

Chapter 18

Flawless Medical Asset Tracking

A Success Story: Challenges and Triumphs

UID Journal asked LCDR Corey Cook, of the USN to comment on their program involving NEMSCON, Navy AIT Project Office and much more. LCDR Cook writes:

In 2004, NEMSCOM in conjunction with the Navy AIT Project Office, embarked on an assessment on Automatic Identification and Data Capture (AIDC) technology to help manage this material. This project reviewed the possible integration of AIDC into medical logistics business processes. NEMSCOM's goal was to facilitate the collection of initial source data, collect and pass the AIDC data to reduce processing times, improve inventory accuracy, increase production efficiency, and enhance "total" asset visibility.

One of the key areas of interest for NEMSCOM focused on surgical instrument identification. At the conclusion of a medical mission, expeditionary hospitals, consisting of tens of thousands of components, equipment, and supplies are rapidly packed and returned to NEMSCOM for cleaning, repackaging, and introduction into the next expeditionary medical platform build. Thousands of surgical instruments, which have no manufacturer markings or means of identification, are returned in boxes, barrels, and footlockers. With more than 14,000 different types of medical instruments, made by a myriad of manufacturers, the process of trying to identify these instruments can be daunting.

The identification process required the skills of numerous senior medical technicians and required countless hours of research with each instrument requiring 15 to 30 minutes for identification. Even with these efforts, identification accuracy was limited to approximately sixty percent. With many of the instruments costing several thousand dollars and having limited availability, identification for future builds was simply not an option.

In an effort to drastically increase medical asset identification NEMSCOM, embarked on a pilot program to develop effortless recognition of medical assets. The marking system needed to be highly accurate, durable, inexpensive, easy-to-operate, and increase production efficiency. The end result of this effort was the implementation of info dots in conjunction with a newly developed software program called MAAT (Material Automated Asset Tracking).

The info dot is a two dimensional data matrix mark, available as a 3 mil and 10 mil bar code which is less than a tenth the size of most common barcodes. This Data Matrix bar code is a small, flexible and unobtrusive label that is virtually indestructible. More than 60 percent of the label can be torn away still allowing for a one hundred percent read rate. It is easily attached to any surface using a pressure-sensitive acrylic adhesive. The info dot has a high degree of redundancy making it highly reliable and its symbology can be read with Charged Couple Device (CCD) scanners. The Info Dot can withstand temperatures of nearly 500°F short-term and nearly 400°F over a long-term period. It resists solvents, caustics, and acids as well as oils, grease, fuels, and salts.

Due to the highly sensitive nature and extreme sterile requirements of surgical instruments, the info dots were exhaustingly tested through rigorous autoclave cleaning to precisely replicate hospital conditions. The end result demonstrated extreme durability, complete sterilization of the instruments and flawless identification of every single instrument.

NEMSCOM also developed a catalog tracking system through its support contractor called the Medical Automated Asset Tracking (MAAT). The DataMatrix bar code links a specific instrument to a data file that contains information about that particular item, which includes, but is not limited to, item location, item identification (NSN), serial number, nomenclature, part number and manufacturer. As new items arrive and are received, they are photographed and the manufacturer data is place into the catalog library.

To date, NEMSCOM has developed a catalog consisting of more than 17,000 items.

Scanning the DataMatrix barcode takes only seconds and an identification of the surgical instrument is certain.

One of the major focuses for the program from the very beginning was to receive a significant return on investment through drastically increased efficiency and total asset visibility. Through the use of info dots and the MAAT system, NEMSCOM completely eliminated the need for costly medical technicians and provided general warehouse personnel the means to identify any item within mere seconds, versus several hours. In one test trial, two senior medical technicians took more than two hours to identify 10 medical items and ended up identifying only four accurately. Utilizing the new info dot technology and the MAAT system, a single warehouse worker identified all ten items in 32 seconds with 100 percent accuracy.

The benefits of this technology have gone far beyond just a warehouse production environment. NEMSCOM has been able to utilize the info dot marking technology to develop surgical tray kitting. Specific surgical instruments can be defined in the MAAT system depending on the surgical procedures required. This drastically reduces the burden on the senior resident nurse allowing any staff member to assemble the tray as required. Once the items have been assembled in the tray, each item can be scanned and verified simply using a laptop with the MAAT software and a hand scanner plugged into the USB port. Following a surgical procedure, each item can be scanned on sight to verify that all items from the procedure were returned, ensuring patient safety and eliminating any potential liability to the staff and the hospital.

As world events dictate an ever increasing presence of the military medical community throughout the world, NEMSCOM will continue to provide the latest technology to ensure the most robust and efficient service and support.

Chapter 19

UID For 'Newbies'

What is it, Who needs it, Why?

Let's break down what all this stuff means. What is UID? Why do I need it if I never did before? Are you sure I need to know this? I understand it has something to do with inventories and getting paid and tracking . . . but what?

Breaking it down

UID stands for **Unique Id**entification. These are common words we all know and use everyday, but take a moment to review the exact meanings of these words:

---

unique 1 : being the only one of its kind : SINGLE, SOLE

identification 1 : an act of identifying : the state of being identified 2 : evidence of identity

    identify -fied; -fying 1 : to regard as identical 2 : ASSOCIATE 3 : to establish the identity of

    identity 1 : sameness of essential character 2 : INDIVIDUALITY 3 : the fact of being the same person or thing as claimed

® 2006 Zane Publishing Inc. The Merriam-Webster Dictionary © 2006 by Merriam-Webster, Incorporate

---

The meaning of 'unique' is straightforward enough, but the definition of 'identification' carries some complexity. Though it may seem pedantic, we call your attention to the three distinctions made for 'identification'—the act, the state and the evidence—because these distinctions are reflected in the implementation of a UID program.

The "act of identifying" can be thought of as occurring during the initial *marking* of an item; while the "state of being identified" is attained when the unique information contained in an item's mark is then *registered* for future reference; and the "evidence of identity" is *validated* when an item's mark is subsequently read and it's *unique identifier* successfully extracted.

Before moving on, you'll notice that the definitions for the roots of identification (identify, identity) contain the words 'identical' and 'sameness' but also include 'individuality'. So, identification actually deals with both aspects; or more precisely, it concerns an item's association with other items both similar and dissimilar. Which explains why you'll encounter the Zen-like requirement for identification "to distinguish an item from all other *like* and *unlike* items." Sometimes you want to identify a specific item for, say, maintenance purposes, while other times you need to identify how many of a certain items you have on hand.

In terms of the Department of Defense (DoD) program under discussion, however, the emphasis remains on correctly conveying an item's individual information, so the modifier 'Unique' is applied to 'Identification"

Some necessary details: watching your 'I's and 'U's

Which leads us to a distinction that may strike you as splitting hairs, but must be made; otherwise there's risk of confusion later.

Notice that the word 'item' has already crept into our discussion (it was hard to keep it out). Besides the now familiar 'UID' there is another term, 'Item Unique Identification' (IUID) which in casual usage is used interchangeably with UID. Technically speaking they are different things.

UID is more general and refers to the overall system of identifying 'entities' and assigning them unique identifiers. The point being, that besides items, the same identification system can be applied such things as personnel and real property. Accordingly, the DoD is also developing parallel UID programs pertaining to these other 'entities.' IUID, on the other hand, refers more specifically to the system of marking *items* that will be delivered to the DoD.

Another acronym full of 'U's and 'I's that you're going to keep encountering is 'UII', which stands for *Unique Item Identifier*. This is the set of data marked on an item that is globally unique, unambiguous and robust enough to be read throughout the lifetime of the item. The UII should not be confused with the physical mark placed on the item. Instead it refers to the information contained in the mark and the word 'set' is used here because this information is complex, containing multiple data elements—the details of which we will return to shortly, but first . . .

Why this whole affair, anyway?

It is helpful to review how this all got started in the first place. Though some might cynically suggest that the DoD intention is merely to make life more complex for it's vendors and suppliers, the truth of the matter is that the DoD is responding to the emergence of increasingly efficient processes and technologies in the private sector for dealing with inventory and it's management. The appropriate response by the DoD is to require that these same methods be brought into it's own organization structures and advanced throughout. This response led to a vision which begat a policy which begat strategic imperatives.

For the UID in general this vision results in the imperative for "DoD, its coalition partners and industry to efficiently and effectively manage people, property and intangible assets using globally unique identification."

Pertaining to the more specific IUID, the vision establishes an imperative to "uniquely identifying tangible items relying to the maximum extent practical on international standards and commercial item markings." The foreseen benefit of the IUID imperative is that "uniquely identified tangible items will facilitate item tracking in DoD business systems and provide reliable and accurate data for management, financial, accountability and asset management purposes."

Along the way, it's important to keep in mind that the DoD vision for IUID is driven by the its logistical need to attain 'in total' the following:

Asset Visibility
Property Accountability

In particular, these goals are stated in the form of objectives to:

Streamline DoD supply chain to better support war fighters
Improve logistics, contracting and financial transactions
Capture the value of DoD purchases
Control item usage
Combat counterfeiting

To attain these goals and objectives, the DoD has implemented a set of policies relating to IUID during the past four years. This process began with the issuing of a Policy Memorandum in July 2003 followed by publication of the Final Rule in April 2005, which was amended June 2005. Most recently, a policy update for IUID of Tangible Property was issued February 2007. More importantly, the deadline for complying with these policies was ~~set for~~ September 2007.

For suppliers to the DoD, the gist of these policies regarding IUID comes down to the following two rules

"Contractors to provide *unique identification* for items delivered to DoD through the use of Item Identification marking"

"Contracts to provide for Identification of the Government's *acquisition cost* of all items built or acquired by the contractor and subsequently delivered to DoD under contract"

As to which items in particular must be uniquely identified by contractors, the policies state the following rules:

All delivered items where the unit acquisition cost is $5000 or more
Items less than $5000 when: identified by the requiring activity as serially managed, mission essential or controlled inventory the requiring activity determines that permanent identification is required

Regardless of value when the items is:

> any DoD serially managed subassembly, component or part embedded within a delivered item the parent item that contains the embedded subassembly, component or part

Still not sure? Then try consulting the following flow diagram:

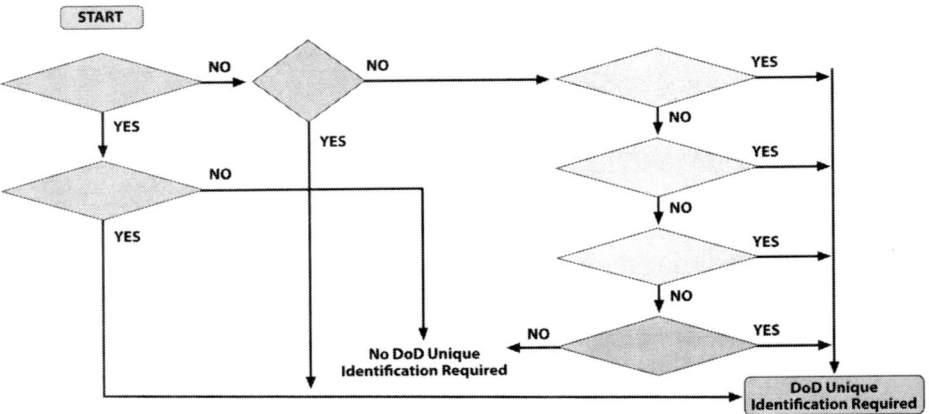

OK, but why me?

Chances are, the reason your participation is required comes from the fact that your company supplies some item to the DoD that must be correctly marked and then successfully tracked throughout its system from "cradle to the grave" And you're somehow involved in the process of assuring that a Unique Item Identifier (UII) gets correctly applied to this item. So let us return now to what constitutes a valid UII.

For good reason, much emphasis is placed on the practical challenges associated with affixing the physical mark to the item—after all it must be legible and last the lifetime of the item. But for now, we call your attention to the logical content encoded into the item's mark, because this information, the UII, is the keystone of unique identification. To attain the DoD's first stated goal of global uniqueness, however, this UII contains some finesse, which is worth elaboration.

As you are no doubt familiar, when an item is manufactured within a company it is assigned a serial number to distinguish it from other similar items. That is, the item is the 10th one produced as opposed to the 100th. The first use of this serial number is to make the item unique within the manufacturing enterprise. This can be attained one of two ways.

One approach is for a company to assure that a given serial number is used only once throughout its enterprise. In a sense, the organization is choosing to ignore the fact that it is making different types of items, so it can use the serial number to assure uniqueness of the item's identification across it's enterprise. We'll call this method 1.

Another approach used by many companies, assures that a given serial number is used only for that type of part. This means, though that two items within the company—though different parts—can have the same serial number. So, to uniquely identify an item within the company, additional information indicating the type of part must be retained and along with the serial number. This additional information, of course, takes the form of a Part number. Furthermore, in some companies an item's serial number is only unique for a given Lot or Batch. In these cases Part, Lot or Batch number information must be combined with the serial number to *construct* an identifier unique to the enterprise. We'll call this method 2.

Accordingly, when the DoD defined the data set for the UII, method 1 is represented as Construct #1 and method 2 becomes Construct #2.

Though closer, we have not yet attained the globally unique identifier. So far, we've only assured that the UII is unique across the enterprise. That is, manufacturer XYZ can have the same number in its UII as manufacturer ABC. The final step is to include an Enterprise identification number for each company and, *viola*, we've achieved the required globally unique UII.

Chapter 20

UID For Newbies—Advanced

Part 2

In the first UID for Newbies, we provided background information on what Unique Identification means and why the UID initiative came to be, while also striving to clarify some of its terminology. What that behind us, we can now concentrate on what you are actually going to be doing—a whole lot of Marking and Reporting.

Defining the Mark If the overall functional goal of Unique Identification (UID) is to construct a globally unique identifier (UII) for a given item (see session one), then the first specific goal of UID in practical application is to successfully affix a physical representation of an item's UII to the item itself—this is the item's Mark, and whole process is called Marking.

Because marking must deal with a broad variety of items composed of various shapes and made of differing materials, it turns out to be one of the trickier tasks in applying UID, so you will encounter a lot of concerns and discussions regarding this operation. Fortunately, to counter these challenges, there is also available a wealth of techniques compiled over time and represented by a host of experienced implementers, vendors, consultants and integrators.

The DoD wanted to assure that its requirements for marking reflected this body of existing expertise, so it settled on an extension of existing bar code scanning technology known as 2D Data Matrix. This emerging standard brings forward all the prior knowledge gained by manufacturers in applying one-dimensional (1D) bar codes to items of diverse materials, but also includes significant improvements, especially when it comes to packing more information into a smaller space.

Use of the 2D Data Matrix has been championed primarily by two industries, the manufacturers of electronic devices, especially printed circuit boards, and manufacturers of automobiles. Both groups were seeking a reliable, cost effective method for uniquely identifying and tracking products through the manufacturing cycle, sales distribution and after-sale warranty verification. Durability is also important because the marks have to survive both severe manufacturing processes and hostile field applications.

Accordingly, both the Electronic Industries Association (EIA) and Automotive Industry Action Group (AIAG) groups settled on the 2D Data Matrix mark as defined in the ISO15415, AS9132, and ISO16022 standards. Similarly, when the DoD specified a part marking technology to be implemented by its suppliers for contractual compliance, it also selected the 2D Data Matrix as called out in the MIL—STD 130 standard.

2D Data Matrix is a two-dimensional matrix symbol consisting of black and white square modules arranged in either a square or rectangular pattern.

Shown in here is a simplified view of a data matrix. Data Matrix encodes information in a machine-readable binary code that is dynamically variable in size, format and density. The coding scheme has a high level of redundancy with the data distributed throughout the symbol. This allows the symbol to be reconstructed even if part of it is missing. The binary code is formed as a matrix. Each binary code symbol has two adjacent sides printed as solid bars, while the remaining adjacent sides are printed as a series of equally spaced square dots. These patterns are used to determine the size, orientation and printing density of the symbol.

Data Matrix codes contain three areas of interest.

When comparing 1D barcodes with 2D barcodes, keep in mind that these are *complementary* technologies—as opposed to *competing* technologies. Many vendors will rightfully continue to use 1D barcodes because they meet their manufacturing needs and helps keep down the cost of item marking. However, when it came time for DoD to choose a standard for marking they were attracted to 2D Data Matrix technology and it's primary benefits:

High information density—bi-directional encoding enables more information to be represented in a smaller area.

Digital encoding—does not require precise sizing of individual elements (cells), because all elements are the same size. The analog encoding used in conventional barcodes means that bar thickness is critical to proper decoding.

Error correction—supports the recovery of encoded information even if the mark is damaged and missing as much as 20% of the symbol.

Secondary benefits that should also be noted, include:

Low contrast readability—mark can often be applied directly to an item even when a low contrast image results, so a separate, higher-contrast label is not required.

Scalability—mark can vary in size (from 0.001 inch up to 14 inches per side) depending on the 'real estate' available on the item for marking. This is limited, of course, by the resolution of the available printing and imaging magnification techniques.

Orientation flexibility—because matrix marks can also be read by video cameras as opposed to a scanned laser beam used to read conventional barcodes, they can be read in a greater variety of orientations.

Error correction turns out to be so important that it has driven refinement of the data matrix standard. So, you may encounter two generations of data matrix symbols. The first subset contains the conventional coding for error correction that was used in the initial installations of data matrix systems. These versions are referenced from ECC-000 to ECC-140. The second subset, referenced as ECC-200, uses the more sophisticated Reed-Solomon error correction algorithms for improved recovery of data from damaged marks. Also, the ECC-200 symbols have a greater number of cells (144) increasing their maximum data capacity to 3116 numeric digits or 2335 alphanumeric characters. Accordingly, the preferred standard is now ECC-200.

Filling the Mark

So far we have been concentrating what the mark *is*—now we turn our attention to what the mark *contains*. This is important because, amongst all the technical details, it's easy to loose sight of the fact that the mark remains a *vehicle* for conveying information. And the key information that ends up being encoded in the data matrix are data structures (elements) that, in turn, contain information to construct the Unique Item Identifier (UII) discussed in session one. Ah, but the process of getting this information into the data matrix turns out to have much finesse, driven by real word needs.

The first of these practicalities comes from the fact that the DoD strives to "minimize the financial and physical impacts" on it's industry partners; so it does not impose unique government data requirements. This means that a manufacturer can choose to use an existing commercial item marking code—presumably one that it has already incurred the expense and time to implement. A commercial identifiers is considered an acceptable UID equivalents by the DoD when it meets the following criteria:

> Must contain an enterprise identifier which is assigned by a registration or controlling authority
> Must uniquely identify an individual item within an enterprise identifier, product or part number

Must have an existing Data Identifier (DI) or Application Identifier (AI) listed in ANSI MH10.8.2, Data Identifier and Application Identifier Standard

The following examples of common equivalent commercial identifiers are given in case your organization has chosen to make use of one:

Global Individual Asset Identifier (GIAI) GS1 *(AI:8004)*
Global Returnable Asset Identifier (GRAI), GS1 *(AI: 8003)*
Vehicle Identification Number (VIN) ISO 3779 *(DI: I)*
Electronic Serial Number (ESN), cellular telephones *(DI: 22S)*

If your company is using one of these equivalents, take note of the *italicized* codes shown for each because this code will be used later to indicate which identifier is in use. The important point with each of these existing identification schemes is that they assure that each item produced by the manufacturer is being assigned a unique identifier.

On the other hand, if your company is not using a commercial identifier (presumably because no such system was in place already), then it is (or soon will be) constructing a unique identifier in conformance with the DoD's "collaborative solution." This solution is defined by the same standard (ANSI MH10.8.2) applied to commercial identifiers, but also cites ISO15434 to cover the actual encoding of the information. More importantly, instructions are given on how to use serial number and other information available within the company to create a UII.

As mentioned in above, two situations must be covered depending on how the company is handling the assignment of serial numbers to items, or 'serialization':

The serial number is unique throughout the enterprise
The serial number is unique only within the original part number

In both cases, the manufacturing company is represented by a unique Enterprise Identifier, which is assigned to the company by a recognized registration authority (e.g. Dun & Bradstreet, Allied Committee, EAN, UCC). The Issuing Agency Code (IAC) represents which authority issued the Enterprise Id. Both of these codes are then placed in data elements to be written to the data matrix; when combined they indicate the identification of the manufacturer. Now that this *globally unique* identifier has been attained for a given manufacturer, the remaining task to attain a globally unique *item* identifier (UII) is to assure that the item can be identified uniquely within the manufacturing organization.

In case one this is simple because the item's serial number is already assured to be unique throughout the manufacturing organization. So, by combining the unique manufacturing information (IAC and Enterprise Id) with the item's serial number the globally unique UII is constructed. Hence known as Construct #1

In case two, because the item's assigned serial number is unique only to the type of part, additional information, in the form of a Part Number, must be used (along with the serial number) to assure that the item can be identified *uniquely* within the manufacturing organization. Though additional Part information must be used the same goal is achieved—that is the construction of a globally unique UII. This is known as Construct #2.

Note that all this UII information ends up being contained in separate data elements and that there is no requirement for it to have it's own data element. Later on, during subsequent reads of the mark, the software of the reader (or system) will *derive* the UII by concatenating information within these separate fields.

Applying the Mark

Although the direct application of data matrix to an item may be preferred by the vendor, the DoD does not dictate this as the only allowed method. The data matrix can be applied in any of three ways—so long as it remains "permanent throughout the life of the item and is not damaged or destroyed in use." Acceptable methods for affixing the data matrix symbol include:

Embedded directly to the item
Engraved on a plate affixed to the item
Attached as a label

Chapter 21

Selecting a Systems Integration Firm

A 'Top-Ten' List of Considerations and Benefits

What factors should be considered when selecting an integration firm may be important over the long term—not merely in achieving compliance but also in ROI.

A partial list of considerations in selecting an integrator:

1) Does the integrator provide a written needs assessment, process evaluation and development plan?

2) Is the integration firm's area of specialty matched to manufacturing unique application?

3) Is the integrator planning to harvest data & plan for the payoff—early on in the planning process?

4) Can the integrator show the benefits of the data to be collected? Data is where you get an ROI: and is application specific. The need for good data-including legacy data—will assure a valuable data harvest.

5) Does the integrator demonstrate a clear understanding of workflow, i.e. integrated part-marking work cell within manufacturing process?

The UID Compliant Implementation Solution should be merged into the manufacturing process.

6) Does the prospective integrator have previous experience with UID Compliant Implementations-with references?

7) Does the prospective integration firm have experience with interfacing with Manufacturing Execution Software and various other company wide software systems, i.e., Oracle, SAP, SQL.

8) Does the integrator provide training and installation, on-site? How about tech support?

9) Does the integration firm provide updates via a Vendor Web Service vs. Local installation of software?

10) Does the integration firm provide ongoing support of part marking templates and documentation library keeping the library updated including a strategy to tracking changes and version #'s-if necessary?

Some additional Benefits of UID Systems Implementation are also worth considering:

Reducing out-of-stocks.

2) Maintenance data and warranty records.
3) Synchronizing data across the supply chain.
4) Tracking work-in-progress (WIP).
5) Tracking parts histories.
6) Improving quality control.
7) Automating business processes.
8) Improving asset utilization.
9) Reducing lead times.
10) Improving supply chain visibility & managing inventory.

Advance consideration of system implementation has been show to yield substantial improvements to the 'bottom-line' based upon the above benefits.

Chapter 22
Ask Dr. UID

Q#1: What's the difference between IUID and UID?

UID stands for Unique Identification, and has mainly been used to refer to Unique Identification of items until recently. Essentially, it is the DNA of parts. DoD is now developing a number of UID initiatives in addition to UID of items. In order to distinguish between these efforts, UID of items is now being referred to as Item Unique Identification, or IUID. Moving forward, "UID" refers to the overarching effort to uniquely identify and link all of DoD's assets, including Items, Real Property, Organizations, Programs, etc. Also, a "UII" is a Unique Item Identifier, which is the globally unique identifier associated with each marked item.

Q#2: How is IUID stipulated or called out in DoD regulations?

In addition to the IUID policy memoranda that have been issued over the past several years, the Defense Federal Acquisition Regulations Supplement, specifically 252.211-7003 Item Identification and Valuation require IUID. The DFARS clause also requires that items be marked in accordance with MIL-STD 130, Identification Marking of U.S. Military Property.

Q#3: We are having difficulty in getting approval from the IUID Registry for uploading our data. We are using WAWF and our executables are in place and functioning okay. Is there a separate process to upload the IUID Registry? Does it require our system to

be tested and approved? What resources are available to assist with registration and approval?

If an organization is already established on WAWF, they do not need separate approval from the IUID Registry to populate. When a DD250 is accepted in WAWF by the government, IUID data will be automatically transmitted to and populate the IUID Registry. If additional functionality such as IUID query capability is desired, then an account would need to be established with the IUID Registry. For web-enabled, non-WAWF data submissions, an account with the IUID Registry would need to be established. This can be done directly from the IUID Registry website.

For electronic, non-WAWF data submissions, an account with the DoD electronic business hub, the GEX, is also required. Further instructions for the data submission, including necessary registration, is outlined on the IUID website at *http://www.acq.osd.mil/dpap/UID/DataSubmission.htm*.

Additional resources for support with issues include:

> IUID Registry: *https://www.bpn.gov/iuid*
> IUID Registry Help Desk: accounts@bpn.gov
> IUID Help Desk: info@uniqueid.org
> WAWF Help Desk: *https://wawf.eb.mil/HelpDesk.html*

Q#4: "What is the difference between Application Identifiers, Data Identifiers, and Text Element Identifiers, and when would I use each of them?"

Over time different industries have developed different semantic formats for encoding data into barcodes. For IUID purposes, there are three formats that are acceptable: AIs, DIs, and TEIs. These were selected because of their common use amongst defense and aerospace industries, and have been developed and managed through international standards bodies. The first question to ask is whether or not your company uses one of the data qualifier formats in its current practice, and if so use that one. For example, if you are in the aerospace industry, you might use TEIs, which are the accepted data qualifiers within the aerospace industry, and are governed by the Air Transport Association standards and listed in their Common Support Data Dictionary.

DIs are maintained and governed by American National Standard (ANS) MH 10.8.2, while AIs are developed and governed by GS1 (formerly EAN-UCC). So if you are a member of GS1, more than likely AIs are the way to go. If you do not currently use any of the formats, then you can choose any of the three. You may want to review the different standards and standards organizations that govern these data qualifiers to determine if any of them make more sense than the others, otherwise, you are free to choose any of them.

Q#5: On shipment pallets of manufactured items bearing IUID, what are the requirements for RFID Labels?

DoD commercial vendors are required to affix passive RFID tags at the case and palletized unit load levels. When multiple RFID Labeled Packages are on a single pallet, an additional RFID pallet label is required. A 'shrink-wrapped' pallet is considered a palletized unit load. The association of the pallet tags and case tags is established in the DD250's/Material Receiving Reports sent to WAWF.

- Office of Assistant Deputy Under Secretary of Defense (SCI)

Office of Assistant Deputy Under Secretary of Defense (Supply Chain Integration) Crystal Gateway

Q#6: When will the IUID's go into full force? Also, how do we enter information, if the contract can be entered into WAWF. It seems that we've only had 5 or so contracts that require these. Often several contracting officers that I've discussed this with don't know what they are. Also, on one of the contracts that do require them, I will not be able to enter the invoice through WAWF because the issuing and receiving agencies are not set up for it. What should I do in this situation?'

The policy is in effect and requiring activities that should be included in the requirements in the contracts. We are working with DAU and the contracting community to develop a continuous learning module for the contracting community to properly address IUID in contracting documents. If a vendor is established on WAWF, then the IUID data will be populated as part of the creation of electronic shipping/receiving documentation. The WAWF website has a training module available for users.

Q#7: One of requiring activities is not yet set up for WAWF, but their contract required IUID's. What am I supposed to do with that data, since I can't enter in the invoice into WAWF?"

Unfortunately, if the requiring activity is not yet able to receive WAWF transactions to support the acceptance of items, then one of the other data submission methods must be used. For WAWF submission, both the vendor and the government accepting activity have to be established on WAWF. If neither, or only one side, is on WAWF, it will not work for IUID data submission. WAWF is the preferred method to register IUID data in the IUID Registry, but until all organizations have established WAWF capability, we will be reliant upon other data submission methods. Electronically, data may be submitted via an X12 Ship Notice/Shipment and Billing Notice (856/857) transaction, an IUID XML transaction, an IUID flat file transaction, or a WAWF IUID Receiving Report/Combo UDF. All four electronic submission methods require access to the GEX. Manually, data may be entered via the IUID Web Entry site.

Q#8: How will DCMA validate property in the possession of contractors if the DD Form 1662 is being phased out?

Once contractors have assigned UIIs to government property in their possession and submitted data to the IUID Registry, DCMA will be able to validate property in the possession of contractors by running reports for the contracts they support from the IUID Registry. The IUID policy update and DD1662 Transition Instructions provide detailed information for the transition process to an electronic GFP submission capability.

Q#9: Will waivers or exceptions to IUID be granted?

The rule is considered to be a strategic imperative, necessary to efficiently move supplies to war fighters. The IUID DFARS final rule outlines the scenarios under which exceptions may occur, including exceptions for (1) items which, as determined by the head of the agency, are to be used to support a contingency operation or to facilitate defense against or recovery from nuclear, biological, chemical, or radiological attack; or (2) a determination and findings has been executed concluding that it is more cost effective for the Government requiring activity to assign, mark, and register the unique item identification after delivery of an item acquired from a small business concern or a commercial item acquired under FAR Part 12 or Part 8. Component Acquisition Executive approval is required for acquisition category (ACAT) 1D programs and the Head of Contracting Activity approval for ACAT II, ACAT III, and non-ACAT programs. These are not exceptions to the IUID requirement itself. Items granted an exception will have to be marked by the Government under legacy implementation efforts.

Q#10: *In regards to the actual validation/scan analysis of the UID, what kind of equipment do you need to validate the quality of the Data Matrix?*

There are two issues with the mark. First, the validity can be determined with a reader—are all the necessary data elements present and are the syntax and semantics used consistent with IUID requirements. Another aspect is the verification of the mark—does it meet the quality standards specified in MIL-STD-130. Vendors provide equipment specifically designed to verify the quality of the mark. This does not mean however that every mark must be verified from the DoD-perspective—we care about getting a quality mark. Process controls and statistical analysis can evaluate whether quality marks are being produced.

Q#11: *Special Cases: What are the requirements for parts marking on cylindrical parts. Is 16% of the Diameter a requirement?*

AS9132 sets forth the 16% limit on diameter of curved surfaces. MIL-STD-130M references AS9132 as the standard for direct part marking, so to the extent that adherence to 130M is required, and then AS9132 would be invoked. Users who have successfully marked items outside of this standard should talk with the customer to determine if it is acceptable to exceed the threshold.

The real issue is with readability. Can the reader see the mark in a "flat enough" projection to actually read all of the matrix. If it goes too far around the circumference, then you don't see all of the mark when the reader takes the "snap-shot". It flattens out what it sees just like pictures of the earth from outer space. They look flat and you can't see things near the horizon as well. Round parts (like tubes) can be marked, but they can't be read if you can't see all of the mark.

Q#12: Please describe the actual UID entry process into WAWF?

The WAWF website provides additional information as well as an on-line training module (at *https://wawf.eb.mil/*). The IUID data would be included in the receiving report submitted into WAWF. Invoices and receiving reports can be submitted by one of three ways: 1) interactive web application, 2) Electronic Data Interchange, or 3) Secure File Transfer Protocol (SFTP).

Q#13: When will suppliers need to mark both UID and RFID on the same contract and how will that work?

RFID requirements are managed separately from IUID requirements. The DFARS clause we reviewed is specific to IUID only. There may be instances where the requiring activity will include contract language/clauses for both IUID and RFID. The best source of information on RFID is the DoD RFID website at *http://www.acq.osd.mil/log/rfid/index.htm*. From the website: "The DoD will not require suppliers to apply passive RFID tags to the packaging of UID items during the 2007 calendar year. The Department will continue to evaluate the appropriate time frame to begin tagging at the packaging level for UID items and will promulgate this requirement in advance of future issuances.

To get your questions answered, please go to our website and submit them: *http://www.uidjournal.com*

Further Resources: *UID Journal AIDC Technology Course*

This online course (TrainingCenter.uidjournal.com) covers all of the aspects of Automatic Identification and Data Capture (AIDC) Technologies. Course presentations are through lectures, equipment labs, system design workshops and self focused exercises. The AIDC course is self-paced and divided into three separate courses or segments:

- Segment 1 concentrates on bar code technologies and the DoD requirement for Unique Identification (UID) on serial controlled assets
- Segment 2 covers RFID and how it is used throughout industry to increase productivity
- Segment 3 combines bar code and RFID of the previous segments to complete system design applications

Course One: Linear bar code, two-dimensional (2D) symbol technologies and Biometrics are introduced and explained so the student has an in depth understanding of how they work. This will help the student apply the appropriate technology for a particular application or environment. Standards and application guidelines for the technologies are explained so the student can ensure that the implementations are compliant. The student will learn about the range of printing and scanning equipment and the key techniques for their implementation, including verification techniques. UID, IUID and UII are explained in detail, including who needs to provide this identification, how it is structured and its proper implementation. All the concepts and techniques associated with the UID are explained, covering the usage of Data Identifiers, Application Identifiers and Text Element Identifiers, as well as metadata requirements. The UID Registry is also explained.

Course Two: Radio Frequency Identification or RFID is the hot new AIDC technology today! It is rapidly being implemented in supply chains and business processes, due principally to Wal-Mart and U.S. Department of Defense mandates for its use. In order for you to be ahead of many people in industry, this segment will provide complete information on what you need to know to design and implement RFID systems to improve business processes and comply with industry mandates. Topics to be covered include RFID standards, physics, and creation of EPC and ISO tags, smart label printers with associated lab. RFID readers with labs, site surveys from a consulting prospective, applications and case studies.

Course Three: AIDC implementation can have a major impact on the entire organization. The system design and integration segment provides a step by step approach to successfully implement an AIDC project without pitfalls. It includes formation of the project team, training, implementation stages, vendor selection and auditing the implementation. Multiple alternatives will be given in the design of an inventory, shop floor, retail and an asset management system. These alternatives will be enforced with multiple case studies. At the end of each session you will have the opportunity to design a system utilizing both Bar Code and RFID technologies. This session will conclude with an inventory workshop where you will put into practice everything you have learned throughout the course.

Students may elect to receive three optional College Credits by completing all three segments. This course is unique because students not only learn the technology and applications from the lectures, but the concepts are reinforced through the self-focused exercises, system design workshops and labs. The on-line class provides equipment labs through videos. Software will be provided allowing each student to design UII direct part marking, printing of labels and data plates. A verifier video lab will then be used to demonstrate verification of the UII mark in accordance to MIL-STD-130. Students will also be able to design and test their own Asset Management system. Further information available at *http://www.uidjournal.com*; *www.uidlabels.co.uk*

Lightning Source UK Ltd.
Milton Keynes UK
24 January 2011
166259UK00002B/150/P